MANY MOONS

MANY MOONS

The Myth and Magic, Fact and Fantasy
Of Our Nearest Heavenly Body

DIANA BRUETON

PRENTICE
HALL
PRESS

New York • London • Toronto • Sydney • Tokyo • Singapore

Prentice Hall Press
15 Columbus Circle
New York, NY 10023

Library of Congress Cataloging in Publication Data
Brueton, Diana
Many Moons: the myth and magic, fact and fantasy
of our nearest heavenly body/Diana Brueton.

p.   cm.
Includes bibliographical references and index.
ISBN 0–13–553322–X

1. Moon – Popular works.
I. Title.
QB581.9.B78 1991
523.3 – dc20

Art direction and design by Jeff Peers of Impressive Graphics
Picture Research by Patricia McCarver
Typesetting by The Word Shop, Bury, Britain
Printed and bound by Cronion S.A., Barcelona, Spain

10 9 8 7 6 5 4 3 2 1
First edition

"The wise man looks not at the finger, but at the Moon
to which it points."

Contents

Author's Preface

There is a curious imbalance in our need and appreciation of the moon, especially recently. It is almost as though this single, seemingly dead world, reflects our every mood here on Earth and therefore lives to a far greater degree than we might imagine. We are interested in her right now both from the scientific viewpoint and the magical, and as such she represents both the past and the present, maybe even the future, for mankind.

Always, she has reflected our earthly fascination for mystery, fear and magic. This is so familiar to mankind that it resides almost in our bones and we are touched by the changes in the moon's cycles and positions unconsciously. But in this century, science has begun to take a greater interest in what has been observed on the moon, and subsequently brought back from her surface. This has satisfied our scientists as well as our mystics. We might even suppose that the scientific realization of lunar presence is merely this century's incarnation of moon-madness. Every age has a method of looking at the surrounding worlds – science is the method of the 20th century. It might appear to be a duality – science and religion – but perhaps it is the same thing in different guises.

The contents of the coming pages, therefore, reflect this so-called duality – a duality which may actually be in accord one day. With a visual and a literary appreciation of our nearest heavenly body in all its aspects, perhaps the reader will appreciate just how alive the moon is and just how much we need her as a reflection of our own planet.

Introduction
by Colonel James Irwin

We were over the moon for three days and then on the surface for three more days in the summer of 1971. Our base for exploration was named Hadley Base, after the British astronomer who developed the sextant that would permit man to navigate on uncharted waters. I was the lunar module pilot of Apollo 15. Some say that I am "over the moon" because of the emphasis I place on my journey to another world. Some would even say that I am a "lunatic" in the most literal sense. One of Webster's definitions of that word is "extravagant folly". Certainly it is very appropriate for me to write the forward for this book, for with that lunatic quality and the direct experience of the moon I am well qualified to comment on our beautiful "night light".

When I was very young growing up in Pittsburgh, PA, I felt even then the pull of the moon. I dreamed of going there. I told my parents of my dream and even told our neighbors that someday I would journey to the moon. You can imagine their reaction. They all laughed at me in 1935. My mother said, "Son, that is ridiculous. Man will never be able to do that, I want you to do something worthwhile with your life". My early dream was squashed.

Many years later I was selected as an astronaut for the Apollo program. We trained for five years for that mission. We studied everything we could about the moon because we were to explore the highest mountains there; the Apennine Mountains that form the nose of the "man in the moon".

The journey to the moon greatly changed my life in every way. There was a physical change and that was well documented by the medical community. The body quickly adjusts to the conditions of a weightless environment. There is a reduction of body fluids and less production of red blood cells, less calcium production, deconditioning of the heart, atrophy of muscles and an intense radiation environment.

The moon is the ultimate desert. Existence there is very precarious. The space suit was regarded as our moon cocoon. Our physical condition was even worse because we became dehydrated on the moon. We worked too hard. We had not selected sufficient cooling in our liquid cooled underwear so we became dehydrated with no replacement electrolytes available. When we left the moon's surface our hearts beat very irregularly.

The psychological changes are not well documented. Our concept of ourselves is largely based on our perception of how other people regard us. Now even my closest friends regard me differently. They don't introduce me as simply Jim Irwin, but 'Jim Irwin who has been on the moon'. Some even think that I should have new senses and appearances because of this. Then there is the grand spiritual change of seeing the earth as God must see it. We lived on another world that was completely "ours" for three days. It must have been very much like the feelings of Adam and Eve when the Earth was "theirs". Yes, our lives were greatly changed.

As you know, there have been twenty-four Americans to travel to the moon. Twelve of these have walked on the moon's surface. Six have had the convenience of the lunar rover. If we had not had the car, we would have been unable to reach the mountains of the moon. We are approaching the 20th Anniversary of Apollo 15. Three Command Pilots have died. Jack Swigert, Apollo 13, died from leukemia. Don Eisele, Apollo 7, died from a heart attack. Ron Evans, Apollo 17, also died from a heart attack. We are a vanishing breed. It won't be long until all of us are gone and there will no longer be any firsthand knowledge of the moon visits. More people may regard the adventure as a myth or a hoax! As long as I am here, I want to share the mysteries of the moon.

One of man's most incredible desires – to reach the Moon – was fulfilled within our lifetime. Now, just a few of the men who ventured there remain to tell the tale.

The moon is ¼ the size of earth. The gravity is ⅙ that of earth. We orbited the moon at 4,000 miles an hour so it took two hours to complete our journey around the moon. The earth appeared as a beautiful blue marble in the blackness of space. I could hold the earth between my fingers or cover it completely with my thumb. Most are surprised when I tell them how small the earth appeared. Have you ever tried to hold the moon between your fingers? If you do, you will find it the size of a small pea. It is ½ degree. The earth is 4 times larger so it appeared as a marble. I could see no evidence of life on the blue planet. I could not see the great cities such as London or New York. I could not see the great construction projects such as the Great Wall of China. But I knew that the blue planet was my home. It is the only natural home for man in the universe as far as we know. The moon was a great contrast to earth. On the moon there is no life, no sound, and no sky.

We looked up into the blackness of space and could not see stars because of reflected light from the sun. It is light for 14 days on the moon and then darkness for 14 days.

On the moon, our existence is termed in superlatives! We were very light because gravity is only ⅙ of earth. The sun was extremely bright because of no atmosphere to reduce the intensity. We had pure light from the sun. We were very hot because of this heat source in the heavens. When the sun is overhead the temperature is about 250 F. We were only on the moon during the early morning hours of the moon day so we were only exposed to temperatures of 180 F. We needed our liquid cooled underwear.

We returned with new knowledge of our nearest heavenly body. We established a network of scientific stations on another world. We have returned 800 lbs of lunar material to be studied by scientists so we will know more about the moon.

Apollo was the Greek God of light. I hope that we have returned with new illumination, new vision and new hope. When shall we return to the moon? Our President predicted that it would probably be 2020 AD. That will mean there will be 50 years between visits to the moon. Perhaps you will be on that next mission! Is that your dream?

My good friend in school was John Young, the only man to fly in space six times. I never dreamed that someday John would fly in space and I am sure he never dreamed I would. But two men from the same school ended up on the moon in the same order that we were in school. I was the 8th man on the moon and John was the 9th man. Isn't that remarkable?

It is a very special time. It is "once in a blue moon"! I hope that you will enjoy *Many Moons* as you gain new knowledge about our moon and its effect upon our lives.

Apollo, Greek god of light and the Sun, lent his name to the first space missions to reach his sister planet, the Moon.

MUTUS LIBER, IN QUO TAMEN
tota Philosophia herme tica, figuris hieroglyphicis
depingitur, ter optimo maximo Deo misericordi
consecratus, solisque filiis artis dedicatus,
authore cuius nomen est Altus.
21. ii. 82. Neg:
93. 82. 72. Neg:
82. 81. 33. Tued.

Part I

MOON MYSTERIES

Chapter 1

Shooting at the Moon

Ever curious, man has looked to the skies, seen the mysterious realms of the stars and planets, and wondered. What is up there? How far away are those shining spheres? Are there any other worlds, might there even be other forms of life out there? He has seen birds winging their way effortlessly through that same space; perhaps even on their way to those distant places. And if the birds could travel in such a way, then why not man, with his superior intelligence?

In early times it would have been quite reasonable to assume that the Moon was just a hop and a jump away – quite possible for a child to believe that a cow could jump over it, or that it might get caught in the branches of a tall tree. Man could speculate that he might reach it, if only he could find a way.

So it is not surprising that man has long fantasized about flying off to other planets. Legends like the Greek myth of Daedalus and Icarus tell of these dreams of flight, as do images such as flying dragons, and Pegasus the winged horse. The idea of traveling to the Moon is ancient. Way back in AD 150, the Greek writer Lucian told of Icaromenippus, who tied the wing of an eagle and the wing of a vulture to his shoulders and flew to the Moon. He discovered that from there humans looked no bigger than ants!

Such is the stuff of dreams; but from early times man has heeded the harsh realities, and has made a science out of studying the heavenly bodies. The Babylonians kept records of all kinds of astronomy as far back as 750 BC. An eclipse of 721 BC is the most ancient reliable Babylonian astronomical observation, suggesting that such an event was considered then already to be of scientific importance.

The Greeks added an immense amount to man's knowledge of the Moon – and, indeed, of astronomy generally. It was they who recognized that the Moon shone by reflecting the Sun's light and that only half its surface could be lit by the Sun. They even figured out that the changing phases of the Moon were the result of its changing position in relation to the Sun. They also came up with a pretty good approximation of the Moon's distance from Earth. In about 270 BC, Aristarchus of Samos came close to the true figure, and by 150 BC Hipparchus of Rhodes was just about spot on as to the actual distance of nearly a quarter of a million miles – a remarkable estimate for so long ago!

Galileo Galilei's discovery that the Moon was a world in its own right, not just a shiny ball in the sky, brought the crowds flocking to peer through his telescope.

So man's perceptions of his place in relation to the moon took a quantum leap. All manner of new possibilities arose from these early discoveries. It must certainly have been at least as exciting as today's scientific findings. Surely, they reasoned, if the Moon was a world, it must contain life, with nations, plants, animals, its own culture, and so on.

And so began the tales of flight to the Moon. As early as AD 150 the Greek writer Lucian of Samosata was sending his heroes there.

Literature, drama and poetry reflected the fashion for this new lunar world. The trend didn't last, however, largely because of the vehement impositions of Christian dogma – one of history's great imagination dampers. Within the established Church doctrine the world was seen as the center of God's creation. So the Moon stopped being seen – for many centuries – as an earth-like world and was mainly regarded, instead, as part of the divine heavens, as in Dante's *Divine Comedy*. Copernicus was the next to stand up to the dull renderings of Christianity, when in 1543 he put forward the "outrageous" proposal that the weighty Earth was actually flying merrily through space, encircling the Sun. What an idea! The Christian planet was no longer the center of the universe, as a result of which imagination could return to enhance all those wonderful fantasies about life on other planets. Poor Copernicus, though, suffered the consequences of the repressive Christians.

A further boost to imagining life on the Moon came when Galileo Galilei turned his telescope skywards. His discovery was one of the most astounding in history, for he found that the Moon was not a smooth, shiny, perfect sphere, but more like the Earth in its uneven, rough surface with a terrain that even included mountains. The Moon has been recognized as a world ever since. The concept of a lump of cheese, or part of a verse in which cows jump over it, died at this point in history and the imagination traversed to looking at our closest neighbor as a real world.

> "... his ponderous shield
> Ethereal temper, massy, large and round,
> Behind him cast; the broad circumference
> Hung on his shoulders like the Moon, whose orb
> Through Optic Glass the Tuscan Artist views
> At Ev'ning from the top of Fiesole,
> Or in Valdarno, to descry new Lands,
> Rivers or Mountains in her spotty Globe."[1]

No Moon-Air, No Moon-Body

Thus dawned the golden age of Moon flight – at least in the imagination! The realization that the Moon had at least some of the Earth's characteristics allowed the imagination to soar, and some dizzy heights were reached.

Today we might scoff at stories like that of a certain Domigo Gonsales, who, according to the highest references in the shape of a seventeenth-century

Bishop of Hereford, had journeyed to the Moon by tying himself to the wings of wild swans, finding the journey easy once he was beyond the clouds! The Bishop believed, along with his local human flock, no doubt, that this particular breed of swans migrated to the Moon in autumn. So romantic a notion was backed up by the understanding, fifty years before Isaac Newton explained gravity, that gravitational pull steadily weakened as one rose through the clouds. And it wasn't until several years after the Bishop of Hereford's story that it was realized that Earth's atmosphere did not extend all the way to the Moon. Until then, Moon-flight writers could happily assume that beyond the clouds there was enough air to breathe – extending the sweet oxygen mix even to the moon itself. If you lived on earth at that time, you

Copernicus shocked the Christian world by showing that the Earth is not the center of the universe – but his discovery gave scope for fantasies of life on other planets.

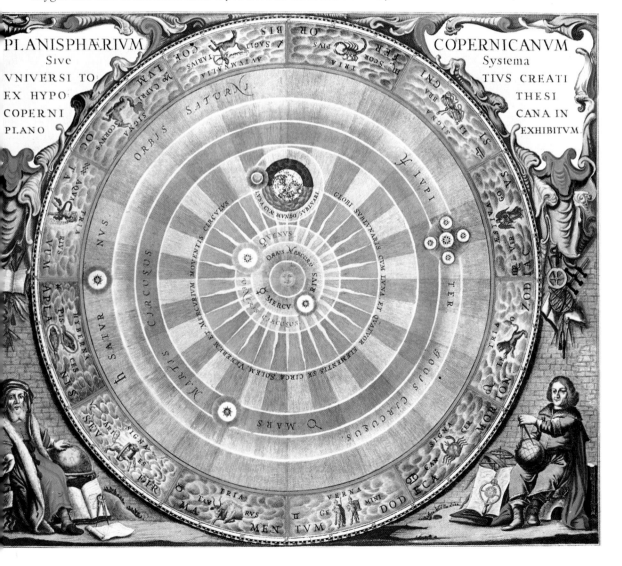

would have no reason to suppose that the conditions you lived with were not universal. Even a seventeenth-century Master of Trinity College, Cambridge, thought that such a journey could indeed be made by tying on wings or using a flying chariot, "in which a man may sit and give such motion unto it as shall convey him through the air".

This happy state of ignorance was being changed by observations through the telescope. It became obvious that the Moon could not possess an atmosphere, because its surface was seen to be always clear and sharp as long as the Earth's was. No clouds or mists ever obscured its features, and the boundary between its lit and unlit part was always sharp, not fuzzed by hazes. The evidence mounted.

What were the implications of this unromantic discovery? Simply, that if there is no atmosphere there is no life, because there could be no air and any sea would boil away under the Sun's heat. So, was this the end of life-on-the-Moon fantasies?

Certainly not! It was "literally" inconceivable to writers that a creation so near and so full of potential could be so useless. The western world was deep into the universal concept of "purpose" as a basis for everything, that nothing should be allowed to go to waste (incidentally not such a bad idea when you consider it), so what could be more purposeful than a richly inhabited planet not so far away from this one? And so the myths continued, unabated by scientific contradiction.

The Discovery of the New World by an eminent Oxford mathematician, John Wilkins, was published in 1640 and set out to prove that the Moon did indeed have an atmosphere, because its sunlit portion seemed larger than its darkened portion. This being so, he reasoned, there could be life on the Moon. In fact, his theory was based on an optical illusion, and his conclusions were therefore pure moon-shine.

Meanwhile, Cyrano de Bergerac (he of the big nose – brilliant French poet, warrior and star of the recent movie) was taking things a little more lightheartedly, although actually coming up with some relatively sound schemes. In his book *The Comical History of the States and Empires of the World of the Moon* (1656) he suggests traveling to the Moon by attaching bottles of dew to the traveller. As the Sun evaporated the dew, he reasoned, the aspiring astronaut would also find himself drifting up towards the Moon. And what we notice about all these methods is their intrinsic poetry, the romantic basis for understanding, largely lost today in the miasma of "hard" science. The writers would create almost any story, any device for getting to the moon, the basis being simply to entertain.

But Cyrano was not totally off the mark – he also suggested using rockets for space travel, perhaps as much with fun in mind as his evaporating technique – an amusing notion – rockets! Three centuries after Cyrano we are into rockets with a vengeance, both for good and bad results.

But rocket propulsion did not hold Cyrano's attention for long. He soon abandoned the scheme in favor of throwing a magnet up in the air while

Above: Earthlings have gazed at the Moon, wondered if it might be within reach and (left) told wondrous tales of venturing into space.

standing on a light iron chariot which would then be pulled upwards. When the chariot reached the magnet it would be thrown again, and so on until the Moon was reached! Unfortunately, this goes against the second law of thermodynamics, but as the latter was not understood for another two centuries, Cyrano certainly could not be equalled in his creative thinking.

But then Newton came along and put an end to all that stuff. It became clear that the Earth's gravitational pull extended a long way and that almost all the space between the Earth and the Moon is a vacuum – quite apart from the Moon having no atmosphere. Once again science attempted to put a damper (along with Christianity) upon the imaginations of the fictional writers. Why would anyone want to visit so barren a place?

For a while, writers used the Moon as a vehicle for social satire.

Swift's *Gulliver's Travels*, a classic of the genre, appeared around this time. In the same vein, *A Voyage to Cacklogallinah* hit the early eighteenth century bestseller list. Written by a Captain Samuel Brunt (probably a pseudonym for a clergyman), it recounts the adventures of the aforementioned gentleman, who is shipwrecked on the island of the Cacklogallinians – human-sized birds with equally human corruptions.

He flies with them to the Moon, where they find a world in which man's dreams are acted out by man's spirit, which comes to the Moon in his sleep. The people of the Moon, the Selenites (a term later adopted by H.G. Wells) give them a feast, and the Earth people learn that the Selenites live in a world without passion or greed, thinking only of philosophy and religion.

More satire and moralizing appear in *The Life and Astonishing Transactions of John Daniel* (1751), in which the author employs a device which allows the hero to fly to the Moon under his own muscle-power, the first time a writer had proposed such a thing. No more birds pulling chariots, no more early-morning dew! Here again we find a shipwrecked man, this time one who fathers a son who builds a flying machine which takes them – by accident – to the Moon. When they return to Earth they are fired on by frightened Earthlings, but eventually John Daniel, "after a life of fatigue and anxiety, reaches England and ends his days in peace and comfort at his native place, aged Ninety-seven Years".[2] So despite the relatively dull findings of the astronomers, in the popular mind the Moon remained inhabitable for many years. And of course, where there is fantasy there is also potential hoax.

In 1834, John Herschel, an astronomer of note, went to the Cape of Good Hope to study the southern stars, in what was expected to be a fairly academic and unexciting expedition. While he was there, however, an American journalist published a series of articles in the New York *Sun* describing an amazingly powerful new telescope that Herschel had built. Through it, the stories told, he had seen never-before revealed detail on the Moon – including buildings and living beings. The public were totally taken in by this story, and the *Sun* sold in massive quantities until the fraud was revealed. The Herschel telescope was somewhat the 19th century equivalent of this decade's NASA failure, the Hubble telescope.

Science Fiction, The Happy Combination

The age of what we would now call science fiction dawned in the mid-nineteenth century. This was the beginning of what has turned out to be a happy combination of fact and fiction, melding to bring often intense and useful insights by imaginative writers into the often pedestrian world of astronomy and "exo-physics". Edgar Allan Poe, for example, wrote several stories in which he is clearly more interested in using an authentic science background than in making social satire.

In *Hans Phaal – A Tale* (1835) he sends an astronaut to the Moon in a balloon:

"The indentures of its surface were defined to my vision with a most striking and altogether unaccountable distinctness. The entire absence of ocean or sea, and indeed of any lake or river, or body of water whatsoever, struck me, at first glance, as the most extraordinary feature of its geological condition."

Above: "The indentures of its surface were defined to my vision."

Left: Fly me to the Moon: in the 18th century story 'A Voyage to Cacklogallinah', birds take the humans to the planet, where they discover a whole new way of life.

It fell to Jules Verne, though, to popularize science fiction. *From the Earth to the Moon* (1865) is very much a story of the journey itself, rather than of who might or might not inhabit the place. In fact, the travellers do not even land on the Moon; the tale is of space flight and its effects on those who undertake it.

In Verne's *Round the Moon* of 1876,[3] a spacecraft circuits the Moon. On returning to Earth, the astronaut Barbicane says he believes that no life could have been there:

"'If the Moon is not habitable, has she ever been inhabited, Citizen Barbicane?'

'I believe, indeed I affirm, that the Moon has been inhabited by a human race organized like our own; that she has produced animals anatomically formed like the terrestrial animals; but I add that these races, human or animal, have had their day, and are now for ever extinct!'

'Then,' asked Michael, 'the Moon must be older than the Earth?'

'No!' said Barbicane decidedly, 'but a world which has grown old quicker, and whose formation and deformation have been more rapid.'"

Verne's book is scientifically based, according to the most up-to-date knowledge of the time. Happily, though, Verne added his own imagination to the formula and is thus regarded as one of the founding fathers of science fiction:

"'And who can say that the Moon has always been a satellite of the Earth?'

'And who can say,' exclaimed Michael Ardan, 'that the Moon did not exist before the Earth?'"

"Their imaginations carried them away into an indefinite field of hypothesis."

And later . . .

"'Console yourself, Michael,' continued Barbicane, 'for if no orb exists from whence all laws of weight are banished, you are at least going to visit one where it is much less than on the earth.'

'The Moon?'

'Yes, the Moon, on whose surface objects weigh six times less than on the Earth, a phenomenon easy to prove.'

'And we shall feel it?' asked Michael.

'Evidently, as 200lbs will only weigh 30lbs on the surface of the Moon.'

'And our muscular strength will not diminish?'

'Not at all; instead of jumping one yard high, you will rise 18 feet high.'

'But we shall be regular Herculeses in the Moon!' exclaimed Michael.

'Yes,' replied Nicholl; 'for if the height of the Selenites is in proportion to the density of their globe, they will be scarcely a foot high.'

'Lilliputians!' ejaculated Michael; 'I shall play the part of Gulliver. We are going to realize the fable of the giants. This is the advantage of leaving one's own planet and overrunning the solar world.'"

In *From the Earth to the Moon*,[4] Verne launches a craft towards the Moon. He describes the craft which takes them there:

"The entrance into this metallic tower was by a narrow aperture contrived in

Above: Until the 20th century, man could only make up stories of reaching to the Moon. Perhaps we now take for granted the incredible achievement of man lifting himself through space (left: Apollo 17 takes off).

the wall of the cone . . . the travelers could quit their position at pleasure, as soon as they should reach the Moon. Light and view were given by means of four thick lenticular glass scuttles . . . Reservoirs firmly fixed contained water and the necessary provisions; and fire and light were procurable by means of gas, contained in a special reservoir under a pressure of several atmospheres. They had only to turn a tap, and for six hours the gas could light and warm this comfortable vehicle."

Such descriptions may appear quaint to us now, with our sophisticated knowledge of what is required of a spacecraft, but Verne had quite an imagination, and his account of the launch of the craft depicts a scene at which he could only guess:

"Instantly Murchison pressed with his finger the key of the electric battery, restored the current of the fluid, and discharged the spark into the breach of the Columbiad.

An appalling, unearthly report followed instantly, such as can be compared to nothing whatever known, not even to the roar of thunder, or the blast of volcanic explosions! No words can convey the slightest idea of the terrific sound! An immense spout of fire shot up from the bowels of the Earth as from a crater. The earth heaved up, and with great difficulty some few spectators obtained a momentary glimpse of the projectile victoriously clearing the air in the midst of the fiery vapours!"

In H.G.Wells' *The First Men in the Moon* (1901)[5], a mythical substance which he names "cavorite" is awarded the potential to counteract the Earth's gravitational pull, enabling two astronauts to land on the Moon. They discover beings called Selenites:

"They have larger brain cases – much larger, and slenderer bodies, and very short legs. They make gentle noises, and move with organized delibera- tion . . .

And though I am wounded and helpless here, their appearance still gives me hope. They have not shot at me or attempted injury."

When the two men become separated, one of them describes his feelings at finding himself alone in this unknown world:

'Then indeed I was alone.

Over me, around me, closing in on me, embracing me ever nearer, was the Eternal; that which was before the beginning, and that which triumphs over the end; that enormous void in which all light and life and being is but the thin and vanishing splendor of a falling star, the cold, the stillness, the silence – the infinite and final Night of space.' (p.235)

Finally, he meets the leader of the Moon beings, the Grand Lunar:

'. . . by a simultaneous movement of ten thousand respectful heads my attention was directed to the enhaloed supreme intelligence that hovered above me.

At first as I peered in the radiating glow this quintessential brain looked very much like an opaque featureless bladder with dim, undulating ghosts of convolutions writhing visibly within. Then beneath its enormity and just

TYPUS MONTIS
Æ T N Æ
ab Authore
Observati
A: 1637.

above the edge of the throne one saw with a start, minute elfin eyes peering out of the glow . . . the eyes stared down at me with a strange intensity.'"

These tales were the transition between the wild imaginings of latter-day fiction and the new movement which arose out of serious findings within the scientific community and as such formed the beginning of the move towards actually stepping on the moon. For it was the work of Wells and Verne that showed that the prospect of traveling to the Moon could now be taken seriously.

"No words can convey the slightest idea of the terrific sound! An immense spout of fire shot up from the bowels of the Earth as from a crater."

Chapter 2

Many Moons Ago

Lunar – Teller of Tall Tales
Transforming Influences

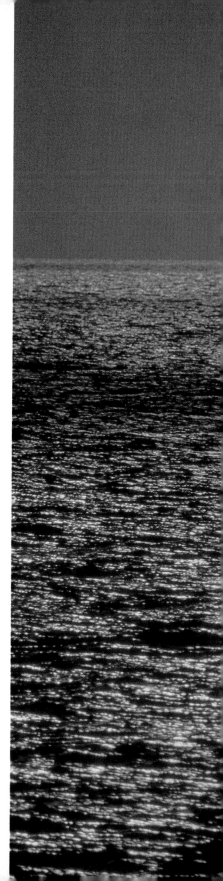

Each month we see the Moon going through its extraordinary, never-ending process of transformation. It appears out of the darkness, a fragile sliver of light, and grows imperceptibly until it reaches the glory of its maturity, then slowly but inevitably fades back into the night. We feel its pull on the waters and life of Earth, and its rhythms seem connected with our own most hidden and mysterious ones – those of fertility, creation, of life itself. Little wonder that myths and legends surrounding the moon are found in all cultures throughout the ages.

In South America, from the Guarani Indians, a tale tells of how brothers Sun and Moon changed into fish so that they could capture the hook and line of an evil devouring spirit. But Moon was eaten by the ogre. The horrified Sun gathered up the fish-bones which the ogre had discarded, and brought his brother back to life. This process of devouring followed by recusitation continues in the Moon's phases.

According to the same sources, the Moon once lived on Earth with his two daughters. One day he saw a beautiful child, and stole its soul which he imprisoned beneath a pot. A shaman was sent to look for the soul, and Moon asked his daughters not to give him away. But the shaman broke all the pots and found the soul. Moon, humiliated, withdrew to the sky with his daughters, to whom he entrusted the task of lighting the way of souls – which is the Milky Way.

The Kororomanna Indians insist that the Moon wanes because it is away hunting. The longer it takes for the Moon to appear, the larger the game he has caught and is busy cooking. So on the Full Moon he is cooking a rat or a mouse, and the size of the game increases daily, rising through porcupine, pigs, a deer, an ant-eater and, on the final day of the last quarter, a tapir.

Still with the South American Indians we come across a sad tale of transformation. Karuetaruyben was so ugly that his wife would not make love with him. One day, as he was sadly pondering on this, Sun and his wife Moon appeared. They were very hairy, with voices like a tapir's.

They asked the Indian to tell them his problem, and to test whether he was telling the truth, the Sun ordered his wife to seduce the man.

But Karuetaruyben was not only ugly, poor man, he was also impotent. So the Sun by his magic powers changed him into an embryo, which he placed in the Moon's womb. Three days later she gave birth to a boy, whom the Sun again transformed, this time into a young man of great beauty. Then he gave him a basket full of fish and told him to return to his village and take another wife, abandoning his old one who had, meanwhile, deceived him.

Much further south in the Wotjobaluk tribe of south-east Australia, the story is told of the time when all animals were men and women. Some died, and the Moon used to say, "You up-again", and they came to life again. There was at that time an old man who said, "Let them remain dead". Then none ever came to life again, except the Moon, which still continues to do so.

The Man in the Moon

An ancient Norse story tells of a man called Mundilfari, who had two children so bright and beautiful that he called the boy Moon and the girl Sun. But this so enraged the gods that the children were taken up to heaven, where the girl became the Sun's coachman and the boy guided the Moon's waxing and waning. Then the boy carried off two more children, Bil and Hjuki, who were carrying water from a well. To this day the children may be seen on the Moon's face. This story is the origin of the nursery rhyme:

"Jack and Jill went up the hill
To fetch a pail of water.
Jack fell down and broke his crown
And Jill came tumbling after."

The significance of the pail of water is that the Moon was believed to influence the waters of the Earth, and to control rainfall, but there is also some Eastern European influence here for the people of the dark regions of Transylvania would always keep a pail of water or a water butt filled to the brim so that passing spirits on their back to the body to become vampires would drown in the water. This was also the origin of the idea that vampires could not see themselves in mirrors because the returned soul would avoid reflection for fear of drowning!

Above: The elusive reflection of the Moon in a pail of water has been used as a symbol of the Moon's influence on us.

Right: Those nursery rhyme favourites, Jack and Jill, were originally Bil and Hjuki in an old Norse legend about the Moon.

The story of Bil and Hjuki's father has also passed into folklore, as the Man in the Moon carrying a bundle of thorns on his back. He is said by Christian lore to have been punished for gathering sticks on a Sunday, and was also spirited away to the Moon.[6]

"A man which stale (stole) sumtyme a birthan of thornis war sett in the moone there forto abide for euere."[7]

"As farre from her thought as the man that the rude people saie is in the moone."[8]

"The man in the Moon was caught in a trap,
For stealing the thorns from another man's gap,
If he had gone by, and let the thorns lie,
He's never been man in the Moon so high."
(Anon)

Primitive man may well have speculated about the distance of the Moon from the Earth, and he most certainly made guesses as to the marks on its face. Some said it was the face of a man. In the Book of Numbers it is said that a man who gathered sticks on the Sabbath, when the children of Israel were in the wilderness, was stoned to death for his sin. This legend somehow became entangled with a pre-Christian one, and the tale that emerged told how the man was thrown up to the Moon for his punishment, where he and his sticks can still be seen with, some add, his dog.

Dying for the Moon

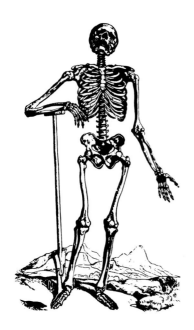

Many cultures associate the Moon with death, but what is most readily noticeable is the position of death as an integral part of life within the stories. There is an ancient belief that the Moon is the home of the dead, either the final resting-place or some kind of half-way house. This may come from the idea of Moon's perpetual life-death cycle. Or perhaps it is the pale, ghostly light of the Moon in the still of the night that has bound the Moon and death together. The Upanishads, sacred Indian texts, say that the Moon is only a temporary resting place before rebirth. Spirits eventually return to Earth through falling rain and, eventually, through man's semen. At the annual Pitcher Fourth festival, a legend is told:

There was once a sister who had seven brothers. It was "Pitcher Fourth" day, on which all women must fast in order to ensure the long life of their husband. The sister was duly fasting, but her youngest brother, taking pity on her in her hunger, climbed a tree and placed a lantern in it, telling her that the Moon had already risen and that she might break her fast. This she did, and immediately her husband died. The woman watched over her husband's body for a year, preventing it from decaying, until the next Pitcher Festival came around. She then cut her finger and allowed blood to flow into her husband's mouth. He returned to life at once, his spirit having returned from the Moon.

The Pitcher Fourth festival today maintains elements of this legend. Wives fast for a day, during which time they meet together and paint a mural of the ancient story, including two moons, and sing devotional songs. When the Moon rises in the evening, each wife finds a place where she can clearly see the Moon, and inscribes a sacred crossroads on the ground. From a spouted pitcher she pours water on to the cross as an offering to the Moon, just as the wife in the tale poured blood into her husband, and just as the Moon is believed to send spirits back to Earth through water and semen. She is, in effect, bringing the spirit of the Moon down to Earth. As a further homage to the Moon she presses food into the mouth of the figures in the mural she has painted, and breaks her fast by serving food to the men of the family.[9]

Death and the Moon are also linked in this story of adultery and vengeance told by the people of Melville Island, North Australia. A man called Purakapali went hunting. His family remained in the camp, and one of his wives, Pima, took the opportunity to go off into the bush with her lover Tjapara, the Moon, leaving her young son alone in the camp.

On his return, Purakapali discovered his son dead. His rage at his wife, and her lover the Moon, knew no bounds when they at last returned.

To appease the grieving father, Tjapara said to him, "Give me your dead son, and in three days' time I will bring him back alive." But Purakapali was too grief-stricken and angry to listen or to trust the Moon, and he fought with Tjapara from one end of Melville Island to the other and killed him.

Taking his dead son in his arms, he walked into the sea and declared, "As my son has died and will never return, so shall all men." And indeed, Purakapali and his son never did return, as, the islanders believe, is the fate of all men. But Tjapara, the Moon, was back again within three days.[10]

The Hottentots of South Africa have their own stories connecting death with the Moon. The Moon told a rabbit to go to Earth and tell everyone that they would be reborn after death, just as the Moon is reborn each month. But the rabbit got the message wrong, and told the people that they would go round just once in life. The Moon, on his return, was so angry at the rabbit's mistake that he hit him with a stick – which is why the rabbit still has a split upper lip. But the rabbit took his revenge; before leaving the Moon, he hit it with his claws, and the marks on the face of the Moon are still clearly visible. There is something delightfully lively about such tales and we can see the same life and death partnership within some of the older parts of the Bible where mention is made of the Moon in connection with the Sabbath, this being the day the Lord rested from his work. The origins of the word probably go back to the ancient Babylonian "shabbatum" (full moon day). Again there is the ancient connection of the Moon with death and resurrection:

A woman wants to get to the prophet Elijah to beg that her dead son's life be restored. Her husband objects: "Wherefore wilt thou go to him today? It is neither New Moon nor Sabbath." (2 Kings, 4)

Above: The Indian Pitcher Fourth Festival brings the Moon and the Earth together through the medium of water.

Below: The word "Moon" probably originates from the Ancient Babylonian.

That old devil Moon . . . the Moon has always been linked with fertility, through potent forces like the devil and carnality (above), or transformed into more "acceptable" versions (left).

The Fertile Moon

Still staying with the moon's power to transform life we can take a look at the old Christian devil, the serpent, and his (or her) association with the Moon, through its ability to slough its skin at regular intervals. The serpent is an ancient and potent symbol, but in Christianity, as with many other positive symbols that the "new" faith turned sour, it took on an evil, lascivious aspect, possibly because of its phallic symbolism and because of the Moon's link with menstruation. The Virgin Mary is often shown treading down the serpent, yet at the same time her connection with the Moon and the mysteries of fertility mean that there is considerable contradiction evident – carnality is put down, but the goddess of all fertility continues to be worshipped. This unreasonable state of affairs probably arose out of the Christian need to compromise with the very powerful earth-god traditions of paganism, so well established within Europe at the middle of the last millennium.

The Moon's association with fertility is ancient, and to be found in many societies. Even today we link the Moon's month with a woman's menstruation cycle, and believe that the full moon is closest to fertility.

The prophet Jeremiah gives a stern warning about such beliefs and customs:

"Seest thou not what they do in the cities of Judah and in the streets of Jerusalem? The children gather wood, and the fathers kindle the fire, and the woman kneed their dough, to make cakes to the queen of heaven."

This latter was probably Ashtoreth, a goddess of fertility with strong lunar connections. Such cakes were also made on the Greek mainland and in Egypt, India and China. Jericho, the city known to us from Bible stories, was dedicated to the Moon goddess Jerah. Even as late as the nineteenth century, in Ribble, Lancashire, cakes were being baked in honor of the fertile Moon. And it has been suggested that Easter's hot cross buns are an old pagan custom associated with Moon worship, and not originally Christian at all.

"The magic bird first came to Earth to bring fertility to man. Then it flew back to its home in the sky, whence, with folded wings, soft brooding, she still watches over the children of men. Mortals call her 'Moon' and sometimes, when people are sleeping, the Moon-Bird floats down from her place in the sky and pecks up grains or other foods. When you look up into the sky on a clear night, plain to be seen are the little star eggs – and how could they get there, if it were not that the great white Moon-Bird had laid them?"[11]

The Uaupe Indians of the Upper Amazon believe that a girl's first menstruation is due to the deflowering of the Moon. Many tribes talk of the Moon roaming the Earth on bright moonlit nights, and his success at making love with each woman is shown by her monthly menstruation. Similarly, in many places, including England, it has sometimes been told that the Moon impregnates a woman, and that the fetus is fed by emanations from the Moon's light. But this child is not perfectly developed, and so is aborted as a "moon-calf", a misshapen embryo which goes into limbo.

Stories once again give us the "true" reasons for our beliefs – such as this one told by the Kuniba of South America, which skilfully explains the associations between the moon and menstruation.

A young Indian woman was visited every night by a strange man. To find out who he was, she rubbed the black juice of the genipa on his face, and next day was horrified to find that her lover was her brother. He was driven out of the family, and executed. Another brother offered shelter to the brother's head, but he became tired with it's demands for food and drink, and drove it out again. So the head rolled to the village, and tried to get back into its hut. The villagers would not allow it, and in desperation the head decided to change itself into something else. It considered becoming water, or stone, but eventually chose to be the Moon, and rose into the sky unrolling a ball of thread behind it. And, to reap revenge on the sister who had denounced him, the Moon inflicted the curse of menstruation on her.

The Cashinawa of the Amazon also claim that the creation of the Moon marks the beginning of a woman's menstruation. The Moon's phases also affect how a child will look. If it is conceived at new moon, the child will be as light as day. If at full moon, it will be as dark as night.

Another South American story tells of rolling heads, blood and lust: Originally the night was completely dark, with no Moon or stars. Then a young girl refused to marry, and was thrown out of the house by her exasperated mother, with the words: "That will teach you not to want to get married!" When the frantic girl beat wildly on the door, the angry mother took

Above: One of life's greatest wonders, conception, has been linked with the Moon in legends throughout the world.

Right: ". . . when people are sleeping, the Moon-Bird floats down from her place in the sky . . . when you look up into the sky on a clear night, plain to be seen are the little star eggs."

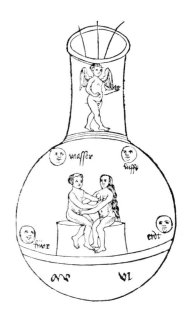

hold of a knife, opened the door and cut off her daughter's head, which rolled to the ground. Her body was thrown into the river.

In the night the head started moaning and rolling around. Realizing it had little future without a body, it decided to change into the Moon, so that it would be unreachable. The head promised the mother that, if she would give it some balls of thread, it would forgive her. It held an end of the thread in its mouth, and vultures towed it up to the sky. Here, the eyes of the head became the stars and its blood the rainbow. And henceforth women would bleed every month.

Men get a Look-in too

If the Moon is linked with women and their fertility, it apparently has even more spectacular effects on men! Many myths connect the Moon with a long penis. This story is told by the Tacana Indians:

A man once caught two thieves, who turned out to be the heavenly sisters, the Moon and the morning star. He fell in love with the Moon, but she spurned him and suggested he pursue her sister instead. In the end she gave in to him, but she insisted that before they made love he should weave himself a large basket. When they copulated, the man's penis grew so long that he had to start carrying it in the basket, where, even coiled like a snake, it still fell over the side.

The man returned to his village, but his troubles were not yet over. At night, his penis would wander off in search of women, and make love with them. One man, whose daughter had been thus served, decided to lie in wait for him. When he saw the wandering member approach, he cut off the end, which promptly turned into a snake. The owner of the now diminished penis died, and the snake became the mother of the termites.

The moon's facility to lengthen the penis also echos amongst the Indians of Bolivia. The Tumupasa believe that the Tapir has a particularly large penis and three testicles because he copulated with his wife just when she had swallowed the waxing Moon, and before waiting for her to release it so that it could begin to wax again.

The African Bambuti pygmies see the Moon as the creator of all things. Everything comes from and returns to *songe abongisi*, the Moon, especially man through menstrual blood, which is also called *songe*. But the Moon also limits our lifetime, because man unnecessarily destroys what *songe abongisi* has created.

In Bambuti cosmology the evening star and the morning star are the wives of the Moon, or his creative forces, or sometimes brothers which build and destroy at once. When man dies, his fire returns to its origin, the Moon, and becomes a child of the Moon.

The Moon, according to legend, even has the power to add a few vital inches to a man's penis!

The Rainbow-Moon

The rainbow is the symbol of the Moon on Earth, and it too has dual characteristics – in the west it is good and portends creation, but if it appears in the east it bodes danger and destruction. In its creative aspect it can cause the eclipses of the Moon, which stop the Moon from killing man. The rainbow's more animate counterparts are the chameleon and the snake. The chameleon lives at the top of trees to be near to his love, the Moon, and like the Moon he is changeable. The snake plays an important part in the initiation ceremonies governed by the Moon, and the Bambuti see the halo of the Moon as a coiled python.[12]

Rainbows have been thought to cause eclipses of the Moon, and to prevent the Moon from doing harm.

Fertile Tortoise!

This potent mix of femininity, fertility, animals and the night is also found in Mexican mythology. Mayauel is goddess of the night sky. She has innumerable breasts, and nourishes the stars who are the fishes of the heavenly ocean. She sits upon a tortoise, who is the beast of the Moon, being able to withdraw into darkness just as the Moon can.

The ancient Egyptians and Chinese also saw the tortoise as in some way connected with the Moon because of its now-you-see-me-now-you-don't ability.[13] The Moon once put on a hat which he could not get off again, according to the North American Salish Indians. He offered to marry the first woman who could get it off. Unlike our familiar fairy tales, this turned out to be the ugly Toad Woman, and from that day on it has been possible for ugly women to marry handsome men.

Animals that Spotted the Moon

How did the Moon get its spots? There are probably as many answers as there are spots.

The Guarani reckon they came about through an incestuous affair by the Moon with an aunt (though nobody mentions where the aunt came from, or why it should have been an aunt in the first place). She stained his face so that he would be recognized. Ever since, the rain falling has been his attempt to wash away the spots.

Alternatively it is said that Beaver and his friend Serpent wanted to marry the Frog sisters. But the Frogs rejected them, considering them too ugly. In revenge, Coyote brought about a flood, and when all the land was covered, the frogs leapt off on to Moon's face, where they can still be seen to this day.

Or perhaps you would prefer this version. Moon invited his neighbors to a great feast. Toad came along but by the time she arrived the house was full and she was turned away. In revenge, Toad made a heavy rain which flooded Moon's house. The escaping guests saw a light coming from Toad's house and sought shelter in the only dry place left. Toad fled, and jumped on to Moon's face, and when people tried to pull her off she clung to it, leaving marks which are still visible.

Wolfsbane and the Moon

Animal stories related to the moon can also, of course, contain a more sinister element, especially with regard to the wolf. It is said that the Moon will one day disappear, and this will mark the end of the world, according to an apocalyptic Norse legend, which contains a strong werewolf flavor.

The Moon is forever fleeing from Hati Hrodvitnisson, a wolf born of a witch, Ironwood, in a wood inhabited by trollwives. One day the wolf will indeed devour the Moon, and then the skies will be sprinkled with blood, the Sun's light will be put out and great winds will arise.

Then the Monstrous Winter will arrive, when brother fights brother, the sea boils and the earth cracks, as predicted in "The Spaewife's Prophecy":

> *Eastward sat the crone*
> *in the Ironwood*
> *Who farrowed there*
> *the brood of Fenrir.*
> *Of their get shall be seen*
> *A certain one*
> *Who shall shark up the moon*
> *Like a shadowy troll.*
>
> *He shall glut his maw*
> *With the flesh of men*
> *and bloody with gore*
> *the home of the gods;*
> *dark grows the sun,*
> *storms rage in summer*
> *Weather's a-widdershins.*[15]

Siriono Moon

Moon was once a great chief who lived on Earth, according to the Siriono of eastern Bolivia. He destroyed a race of evil people, from whom sprang forth the reeds from which the Siriono make their all-important arrows. Moon then created man and the animals, and is responsible for everything in the world. The Siriono tell this story of why Moon retreated to the sky:

Yasi (Moon) had a child. One day the jaguar was delousing the child, and bit it too hard on the head. The child died. Yasi wished to know who had murdered his beloved little one, but none of the animals would give the jaguar away. So Yasi, in his wrath, stretched the neck of the howler monkey, put

spines on the back of the porcupine, twisted the feet of the anteater and threw the tortoise down so hard he could no longer walk fast. That is why the monkey now throws fruit at anyone who passes, in case it is Yasi.

Yasi was still angry, even after this outpouring, and decided to go up into the sky, where he remained, and is still a great chief.

Now, the Moon spends half of his time hunting; when the Moon is dark, he is off hunting in distant places. The explanation of the waxing of the Moon is that when he returns from the hunt his face is very dirty; he washes a little each day until, when the Moon is again full, his face is clean. As for the waning Moon – when he goes hunting, his face gets dirtier each day, until it cannot be seen.

The Siriono also believe that the Moon causes thunder and lightning, by throwing down wild pigs and jaguars from the sky – or, in another version, by pulling up bamboo shoots in the heavens. Yasi also created all the other planets and stars, which are called moon fires.[16]

Legends of Moon and Sun

The Moon and the Sun together form a heavenly partnership. The Moon is often – but not always – seen as the feminine part of this, the intuitive, mysterious creative force, and we see this partnership reflected throughout many world cultures.

The Chinese legend of yin and yang, roughly translatable as the male and female principles, involves the Sun and Moon. And once again we see the story unfold through legend.

Heng O was married to Shen I, a great soldier. While he was away fighting she saw a beam of light coming through the roof, and the house was filled with a delicious smell. She climbed up to the roof, and found the pill of

Thunder and lightning are caused by the Moon throwing wild pigs and jaguars down to Earth, according to a Bolivian legend.

immortality, which she swallowed. Suddenly, she knew she could fly, and was just about to try it out when Shen I appeared. There was no pill for him, and his wife was filled with fear. She opened the window and flew away, but her husband pursued her, and saw her flying towards the full moon. Just when he seemed to be catching up with her a blast of wind sent him plummeting to the ground.

Heng O carried on flying until she reached a huge, cold, shining sphere, like glass. The only living things on it were cinnamon trees.

All at once she coughed, and vomited up the covering of the pill of immortality, which became a pure white jade rabbit. This, says the legend, was the ancestor of yin, the female principle. Heng O decided to stay living in this place – the Moon.

Shen I, meanwhile, was given the Palace of the Sun as reward for his brave fighting. He was also given a lunar talisman, so that he could visit the Moon, and yin and yang might be united. However, it would not work the other way round – Moon would not be able to visit the Sun. For this reason, the light of the Moon comes from the Sun, and the Moon is light or dark according to the Sun's visits.

Shen I did visit the Moon, traveling on a ray of sunlight. When Sheng O saw him she made to run away, but her husband assured her he would do her no harm, and together they built a palace. From then on, he went to visit her there on the fifteenth day of every month. It is the meeting of yin and yang, male and female, which causes the Full Moon. In some stories, Heng O later changed into a toad, whose shape can still be seen on the Moon.

Brother Sun, Sister Moon

Many Eskimo myths tell of the Sun and the Moon being the result of a fight between sister and brother. The brother had sex with his sister, and they were punished by being lifted up into the heavens, where they became Sun and Moon. But which one is which is sometimes variable!

The Netsilik Eskimo, who tell this tale, say that the Moon brings luck to the hunter and fertility to woman. Hence their dire warning that women should not sleep exposed to the Moon, lest they become pregnant. In that climate, it seems quite a sensible precaution for anyone![17]

There is yet more incest in a Brazilian tale of the creation of the Moon. Sex between brother and sister was instigated by the brother, whose face was stained black by his dishonored sister. He fled to the sky to escape his parents' wrath, the sister having become pregnant. The sister was changed into a water bird, or a large animal covered in ticks, while he remained in the skies with his dirty face.[18]

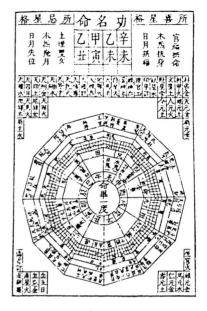

Top: The Chinese have brought together Sun and Moon, yin and yang, in legends and astrology.

Bottom: The potent combination of Sun and Moon has given birth to many tales of sexual adventure – and misadventure!

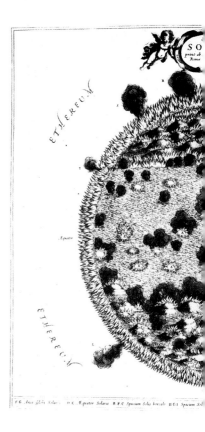

Lunar-Love's Fire

Rather more respectably, an Amazonian story tells how the Sun and the Moon were once engaged. But it seemed impossible for them to marry; the Sun's love would set fire to the Earth, the Moon's tears would flood it. And so they resigned themselves to living apart. But even this had its conditions – for if they were too close together they would create a rotten or a burnt world, and if they were too far apart they would imperil the vital regularity of night and day. And so they must stay forever fixed in their relative distances from each other and the earth.

Buddha's Palm

In Chinese mythology, P'an Ku is said to have created the universe from chaos. He is sometimes shown holding the Sun in one hand and the Moon in the other. But originally he did not place them in the right positions, and so the world was in darkness. Then Terrestrial Time asked that they should be repositioned to bring night and day. Finally Buddha intervened: he wrote the character for Sun in his left hand, and that for Moon in his right, repeated a charm seven times, and Sun and Moon went up to the skies and separated into day and night. "Then said Sun to Moon: 'Now our children are all married. Come, let us go!' 'Yes!' agreed Moon, 'let us go. You shall light up for them by day, and I by night.' Then they assembled all the people in the plaza and Sun spoke: 'My children! Now I am going off with my godchild.' And the

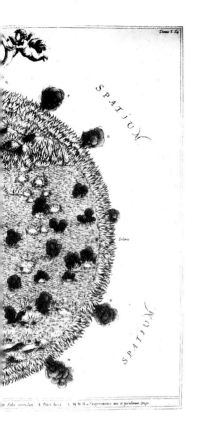

Moon replied, 'Well then, let us go, my godfather!' Then both rose to the sky."
The Kaingang of Brazil tell this story of the origins of the Sun and the Moon. It has a comforting reassurance to it, a kind of parental guardianship felt both day and night, a promise of a perpetual divine presence. The Kaingang go so far as to say that the Sun and the Moon actually created man, by throwing gourds into the water. Sometimes they are seen as antagonistic towards each other, just as two different tribes are, but irrevocably tied together and mutually dependent.[19]

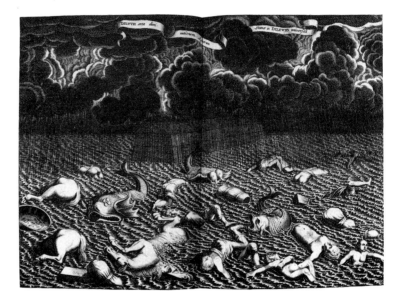

Legends from around the world draw on the Moon: (clockwise from far left) a South American tale describes the love between the Sun and the Moon, whose tears would flood the Earth if they wed. And the Chinese say the world was created from chaos when Buddha sent the Sun and Moon into the sky.

The Benign Moon Still Stained

"Old Woman's Grandchild, son of the Sun by an Indian woman, destroyed the monsters which once infested Earth." So say the Crow Indians of North America, who also believe in the benign influence of the Moon. "Yonder he went. The North Star is he. This woman turned into the Moon. This is the end."[20]

Once Wei and Kapei, the Sun and the Moon, were inseparable friends. So say the Arecuna, who also link the Moon with menstrual blood. At that time Kapei had a pure and graceful face, but things changed when he fell in love with one of the Sun's daughters and began calling on her a little too regularly for Wei's liking. He ordered that his daughter smear her lover's face with menstrual blood. Ever since, the Sun and the Moon have been enemies; the Moon's face is forever stained, and he keeps well out of the way of the Sun.

The Original Moon

All these stories are only a tiny sample of the whole of legend surrounding the moon. A hundred books could be written and we would never come to the end. It is mankind's greatest pleasure to illucidate life through story-telling and traditionally the stories were actually told by word of mouth, not through reading. Or at least, the reading would be done allowed. This sort of effort in relating between, very often, mother or father and child, was a deliberate, though often unconcious act, for now medicine and psychology have begun to realize that the spoken word related to tradition and legend touches a deeper part of the brain than can ever be touched by other methods of learning such as television or visual learning. The voice and the legend combined in love continue the force of life from generation to generation and it is our generations that have for the first time lost this force, hopefully only temporarily, but nevertheless a whole set of our young may grow up without it.

And of course, part of this original learning system related to the original moon. The origin of the moon was, of course, also related to the origin of Earth too. A tale from Indochina explains how the different languages came to be:

At one time all people lived in just one village, and they all spoke the same language. But at a meeting it was agreed that the Moon's phases were blighting their lives – theft and warfare kept occurring at the time of the new Moon – and so they decided to catch the Moon and make it shine all the time. To do this, they started building a huge tower.

It was so high, and took so long to build, that as time went on people started living on the different levels so that they didn't have to keep traveling up and down. After a while each floor began to develop its own language and customs. Then one day the Moon realized what was being plotted. It created a violent storm and dashed the tower to the ground. When the people fell from it, they built new villages where they landed, and carried on their languages

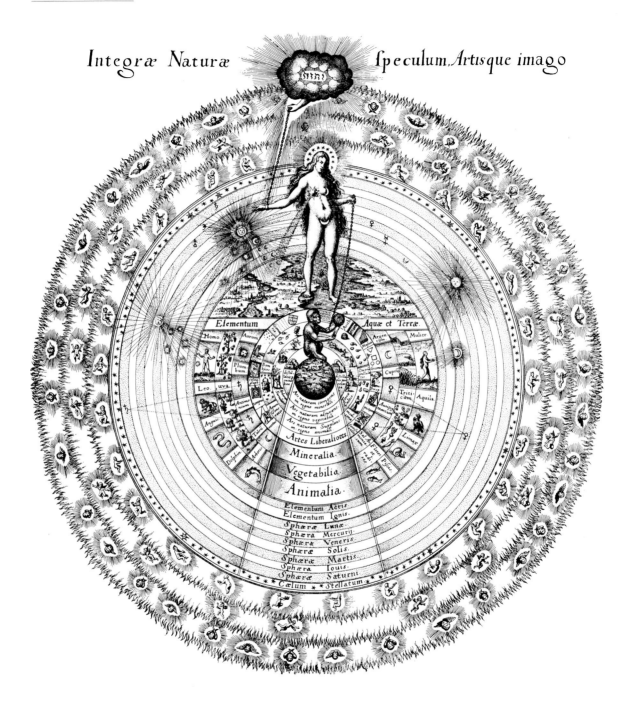

and customs. As for the rubble from the tower, this was said to have become the mountains that run between Burma and the Bay of Bengal.

The Shona people of southern Africa have their own still more violent version of this origin.

The Earth spirit, Mwari, made Mwedzi, the Moon, at the bottom of a pool and gave him a medicine horn. In spite of Mwari's warnings, Mwedzi wanted to go on to dry land, which was barren and lifeless. Mwedzi complained at this, and Mwari gave him a girl, Massassi (Morning Star), to be his wife for two years. She was given tools to make fire. The two made a fire in a cave and lay beside it. Mwedzi dipped his fingers into the medicine horn and touched the girl's body, and she gave birth to grass, bushes and trees, which, when they had grown, made rain to fall.

Mwedzi and Massassi now lived bountifully, but when two years were up Massassi had to return to the pool. Then Mwari gave Morongo (Evening Star) to Mwedzi, again to be his wife for two years. They too anointed themselves from the medicine horn, and she gave birth to animals and, eventually, to human children.

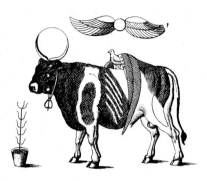

The cow, with its crescent-shaped horns, has often been depicted as an animal dedicated to the Moon.

And so Mwedzi came to be king of a great people, but when drought and famine struck the land the people consulted the divining bones, which advised that the king be sent back to the pool. Mwedzi was strangled, and buried along with Morongo, and remains in the pool to this day.[21]

The idea of the Moon as a chieftain existed also in pagan New Guinea.

Although they believed in many different spirits, the most important were those of the sky, particularly the Sun and Moon. The sky beings looked like humans and always carried torches, the Sun and Moon being borne by the headmen who were responsible for the seasonal cycles, languages and even marriage rules. The natives would set out offerings of food to the heavenly bodies, leaving them for a few hours until they had consumed the "essence". After this the earthly elders would tuck into the physical substance left behind![22]

Once upon a time, say the Yana, night did not exist. There were three families of celestial beings: the two rainbows, father and son, with their uncle who made the rain; Moon and his wife, with their daughters the stars; and Sun and his wife with the meteors, their daughters.

The young Rainbow wished to marry one of the Moon's daughters, Morning Star. But Moon always took his daughters' suitors to be tested by the Sun, which invariably resulted in them dying through some dreadful ordeal. However, Rainbow was helped by his uncle, master of storms, and survived Sun's test. Now Moon was really angry, and tried to destroy Rainbow. But Rainbow was again victorious, and sent Moon, along with his daughters, into the night sky. Here, Moon would die and ever more be periodically reborn.[23]

Left: Man has always tried to make sense of the universe, making up legends and images to give it shape and structure.

Phased by the Moon

To modern human observers the phases of the moon are entirely taken for granted so that, like many other magical aspects of life, we have lost, through science and explanation, the secrets that once existed. Tribes of the past created much more interesting explanations, based on their hearts and their feelings.

An explanation of the Moon's phases comes from the Kapei, who say that the Moon has two wives, one in the east and one in the west. He goes from one to the other; one feeds him well, the other neglects him, so he alternates between fat and thin. This situation can never be resolved, for the women are jealous of each other and will always stay far apart.

Again, we take it without any thought, but the ancients believed in romance. Just because we know now that light and darkness alternate, does not mean it was always so. Some say that once there was only darkness, others that all was once just light. But either way, things needed adjusting if man was to live. The Yupa believe in the all-light version, saying that originally there were two suns. One would rise as soon as the other had set, so daylight was perpetual.

But then along came a woman called Kopecho, who danced around a fire and tried to seduce one of the suns. The Sun got so close to the woman that it fell into the red-hot cinders, and when it emerged it had changed into a moon. Ever since then, day alternates with night. The Moon, incidentally, was furious with Kopecho, and threw her into the water where she became a frog.

Holy Moons

The best-known of the tales of Mohammed's miracles concerns the Moon. Habib the Wise asked Mohammed to prove himself by cutting the Moon in two. Mohammed raised his hands to the sky, and commanded the Moon. It descended to the top of Kaaba (now the stone building at the center of the Great Mosque of Mecca), made seven circuits, then entered Mohammed's right sleeve and came out of the left. Then it entered the collar of his robe, went down the skirt and split in two. One half appeared in the east of the skies, the other in the west, until they reunited.

The Buddhists, on the other hand, say that the Sun and Moon travel along three paths. When they are in the goat path there is no rain, because goats hate water; when they are in the elephant path the rain pours because elephants love it; and when they rise up to the bull path both rain and heat are moderate. They say that the Moon lives in a palace of gem, the outside of which is silver, for both are cold.[24]

Around 2,500 BC, the peoples of the plains of what is now Iraq – a region made fertile by the rivers Tigris and Euphrates and by man-made canals – called the Moon "Sin", the leader of all the sky gods. He gave life, controlled time and drove away sinners, and gave birth to Shamash, the Sun, and Ishtar,

the planet Venus. His divinity, they said, was "like the far-off heavens, fills the wide sea with fear".[25]

The main center of worship was the city of Ur, in which, early this century, the archaeologist Woolley unearthed the remains of a massive tower or ziggurat, built around 2120 BC by King Ur-Nammu and dedicated to the Moon-god. It is fascinating to speculate that Abraham, who was probably born at Ur, might have been witness to the ceremonies around the ziggurat.

The New Moon Goddess

Moon worship, myths and legends are more ancient than almost any other form of hand-down story. As the Sun stands for all things permanent, the Moon symbolizes change and transformation, the mystery of life.

Although early man could not know the scientific explanation of the Moon's waxing and waning, the very fact that it moved through the sky and made this perpetual transformation meant that it must be a god or goddess, with a will of its own. Birth, growth, decay and death are all there, each lunar month, for anyone to see.

In the past few years, especially in the United States, the cult of the Goddess has grown back to a prominent influence, partly through the force of the woman's movement, but also through general dissillusionment both for organized modern religion and also in the practicality and pedestrian nature of science and technology. Both religion and science seem to have done us so much harm recently and mankind searches once again for mystery.

The Goddess movement is most concerned to bring back to the hearth, home, power and influence of women and men, the understanding of how the world and its people used to see the earth, moon and stars. Even the very new male movement, led by writers such as Robert Bly, have touched into that ancient realm of ritual and legend once more, in order to bring back the confidence that has long remained subdued by reason.

Earth used to be the greatest Goddess, Gaia – Mother Earth, bringer of the harvests, provider of all life and center of the cycle of existence, and the Moon stood close by her side, another female, another provider, washing the planet with the tides and helping the harvests to succeed. The pagan beliefs, with their apparent violence of sacrifice, were a powerful signal that mankind was more in touch with the earth than he ever was during the Christian era, and certainly more in touch than he is today.

Chapter 3

Goddess Moon

"Queen and huntress, chaste and fair,
Now the sun is laid to sleep,
Seated in thy silver chair,
State in wonted manner keep:
Hesperus entreats thy light,
Goddess, excellently bright."[26]

I sis, Diana, Selene, Helen, Hathor, Artemis . . . the ancient civilizations created many goddesses whose power and influence came from the Moon. Their essence was that of the Moon itself, with its subtle but vital pull on the forces of life. Little wonder, then, that the Moon goddesses assume many different forms, just as the Moon does.

Some are chaste and virginal, others fecund and richly fertile; some represent the depths of the "other worlds", and of death. They are sometimes to be feared, sometimes to be worshipped, but always to be revered for what they represent – change, transformation, mystery, creation.

Many goddesses have had lunar attributes ascribed to them simply as part of their femininity. Others have been more specifically connected with the Moon; they bring the Moon down to Earth to bestow her gifts on we mortals.

Above: "She is the beautiful eye of night . . ." Selene, Moon Goddess of the Ancient Greeks.

Left: Images of the Moon Goddess have come down through the ages to us, touching us with their beauty just as the Moon does.

Isis, The Embalmer and the Animals

The goddess Isis is known as the Mother of Egypt. She is a beautiful and enduring expression of the Mother Goddess, and was Moon Goddess and Goddess of Waters. The myths surrounding her go back to 3,000 BC, and tell of her precious feminine qualities. She was still being widely worshipped throughout the Roman world as late as the first century AD, though the Roman system demanded less real religiousness and more a practical acceptance of the Goddess to satisfy the local pagan requirements.

The story of Isis tells of how she and her husband-brother Osiris ruled Egypt well and wisely; she taught women the arts of spinning, husbandry and healing. When Osiris was killed by his violent brother Set, her heart was

broken. After much searching she found his body, which Set had thrown out to sea and then choppped up when it was washed ashore. She found all the pieces of the body except the phallus, joined them together and made an artificial phallus, thus performing the first rite of embalming, through which Osiris was restored to immortal life.

The Isis headdress of the disc (the moon) set between two cow's horns, is commonly seen in movie scenes depicting Egyptian rituals such as "Cleopatra". Isis mirrors the Moon's regenerations – her devotee Apuleius described her in *The Golden Ass*:

"I am the natural mother of all things, mistress and governess of all the elements, the initial progeny of worlds, chief of the powers divine, queen of all that are in hell, the principal of them that dwell in heaven, manifested alone and under one form of all the gods and goddesses. At my will the planets of the sky, the wholesome minds of the seas, and the lamentable silences of hell are disposed; my name, my divinity is adored throughout the world, in diverse manner, in variable customs, and by many names."

The power of animals standing alongside the Goddesses is a recurrent theme throughout mythology, for mankind lived with and stayed close in touch with his animal partners. The shamanic priests even dressed as animals and it is said, transformed themselves during trances into the animals they loved.

Ancient Egyptians believed the cat to be an animal of the Moon. The cat-bodied or cat-headed goddess Bast represented the Moon's power over pregnant women. Her son was believed to *be* the Moon, with the function of making women fruitful, so that the human germ might grow in the mother's womb, activated by the Moon's light.

Within Eskimo mythology the Moon is the dwelling place of "Sedna", the Old Woman or Lovely and Glorious Lady. She is the owner of the mammals, and if she is offended by the sins of men she closes up the life-sustaining oceans.

Above: The cat, creature of the night, was linked with the Moon by the Egyptians.

Left: "At my will the planets of the sky, the wholesome winds of the seas, and the lamentable silences of hell are disposed." Isis, Mother of Egypt.

The Mothers of God

The mother of God in all her forms was nearly always that mixture of feminity and fear that mankind is so accustomed to in more earthly women, though of course, exemplified by metaphor and extra powers to act on behalf of the sons and daughters of man.

To the ancient Incas, Mama Qilla – the Moon – was the supreme goddess, governing all things feminine just as the Sun had sovereignty over all matters masculine. She was the object of adoration for women, and even in post-Inca Andes was often asked for help in childbirth and conception. But the Moon – both sister and wife of the Sun – was also to be feared. At the time of an eclipse, out of rage at her own loss, she could turn a woman's spinning tools into vicious animals.[27]

Still more notorious and famous to modern man comes Mary the Madonna, one of the supreme examples of Mother Goddess, and as such incorporating many ancient Sun and Moon beliefs. She is turned to for help with childbirth, crop-growing, healing – the traditional nurturing values. In the early Church, Ambrose and Augustine saw the Sun as Christ and the Moon as the Church. Later, the Moon became more associated with Mary, and she took on all forms of lunar symbolism and imagery. Pope Innocent III told sinners:

"Towards the Moon it is he should look, who is buried in the shades of sin and iniquity. Having lost grace, the day disappears and there is no more sin for him, but the Moon is still on the horizon. Let him address himself to Mary; under her influence thousands every day find their way to God."[28]

Being the Moon Goddess, Mary also had responsibility over the seas and tides, for, in any event, her name derived from the Latin for sea, mare. Certainly, she is always shown wearing blue, which represents both sea and sky.

The Classic Myths of the Goddess

"The mighty hunter, lifting up his eyes
Towards the crescent Moon, with grateful heart
Called on the lovely Wanderer who bestowed
That timely light to share his joyous sport;
And hence a beaming goddess with her nymphs
Across the lawn and through the darksome grove . . .
Swept in the storm of chase, as moon and stars
Glance rapidly along the clouded heaven
When winds are blowing strong."
(William Wordsworth, "Excursion")

The Moon was given many different names and personae within classical mythology. She was Hecate before rising and setting, Astarte when a crescent, Diana or Cynthia when higher in the sky, Phoebe when regarded as the sister of Phoebus, the Sun. She was personified as Selene or Luna, the lover of the sleeping Endymion.

As Selene, she is the lovely goddess of the Moon in the heavens:

"She is the beautiful eye of night . . . Like the sun, she moves across the heaven in a chariot drawn by white horses from which her soft light streams down to Earth, or she is the huntress . . . she is the bride of Zeus . . . the full orb which gleams in the night sky . . . she is beloved by Pan, who entices her into the dark woods under the guise of a snow-white ram . . . The soft whispering wind, driving before it the shining fleecy clouds, draws the Moon onwards into the sombre groves."[29]

Selene bore Endymion, a shepherd, no less than fifty children, and the Moon's passage across the sky was Selene on her way to her sleeping lover, "the Moon Goddess pausing in her nightly course across the heavens to stop and kiss the sleeping Endymion, the setting Sun." (William White, quoted in

Left: The Virgin Mary took on many characteristics of the old Moon Goddesses, such as protector of childbirth and healer.

Below: Endymion, sleeping lover of Selene, is an allegory for the Sun shining life-giving light on the Moon.

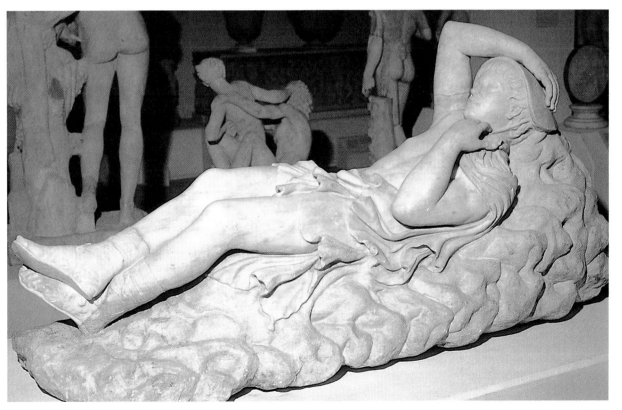

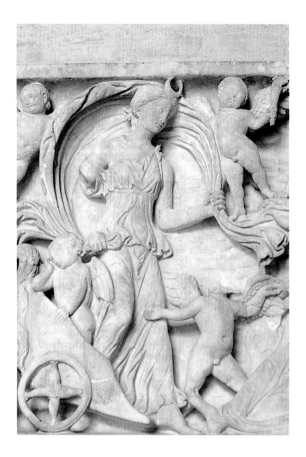

P. Katzeff, op.cit.). One story says that Zeus gave eternal life and youth to Endymion, and Selene came down to Earth every night to embrace him.

"The moon sleeps with Endymion,
And would not be awakened."[30]

"Lo! She rises crescented! He looks, 'tis she,
His very goddess: good-bye earth, and sea,
And air, and pains, and care, and suffering;
Goodbye to all but love! Then doth he spring
Towards her, and awakes . . .
And Phoebe bends towards him crescented."[31]

In ancient Greece cloven-hooved animals were considered to be dedicated to Selene, because their hooves looked like two back-to-back "Cs", the symbol for Selene in the earliest Greek script. The sign also looks like the two halves of the lunar month. So hooved animals were sacrificed at New Moon festivals, with Selene's sign branded on their flank.

Semele was an early Greek version of Selene. During the Festival of the Wild Woman in Athens, a bull would be cut into nine pieces and sacrificed to her – one piece being burned, the rest eaten raw by worshipers. Nine Moon-

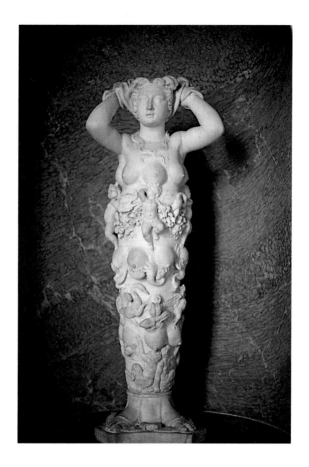

priestesses took part in the feast. And in Greek mythology, the first Labor of
Hercules was to kill and flay the Nemean lion, a huge animal who could
withstand metal and stone weapons. This fearsome beast was said to have
been borne by Selene, who, "with a fearful shudder dropped it to earth on
Mount Tretus . . . in punishment for an unfulfilled sacrifice, she set it to prey
upon her own people."[32]

The irrepressible Pan is also associated with Selene. He once seduced her by
disguising his goatish appearance with snowy white fleeces. Selene agreed to
ride on his back, and to let him do with her as he wished. This probably refers
to a moonlight orgy which was held on May Eve. The young Queen of the
May would ride on her man's back before celebrating a "greenwood" marriage
with him.

So closely has this ancient goddess been associated with the Moon that she
has given her name to the science of the Moon: selenology.

The Confusions of Artemis

Alongside Selene in classical mythology are the Roman Artemis and the
Greek Diana. The tales are quite confusing at times, one goddess merging into

another as worshipers moved around, taking their own particular favorites with them. Artemis and Diana have many common characteristics, and it is sometimes hard to distinguish one from the other. They are both sometimes known as the huntress, "chaste and fair", the virginal girl with the silver bow of the new moon. Artemis was the twin sister of Apollo, and they both had the power to send sudden death or to heal. She protected small children and all sucking animals, but she also loved hunting, especially stags. When she was a three-year-old child, her father Zeus asked her what she wanted. She asked for eternal virginity, a bow and arrow like Apollo's, the ability to bring light, a saffron hunting-robe with a red hem, and sixty young ocean nymphs and twenty river nymphs to feed her hounds. All these she was granted. The silver bow she carries stands for the new moon.

Artemis sat on a throne of pure silver covered in a wolfskin and shaped like a new moon. She hated the idea of marriage, although she delighted in caring for mothers of new born babies. She loved hunting and fishing, and swimming in moonlit ponds. Should a mortal see her naked, she would change him into a stag and hunt him to death. Despite her virginal ways, though, at one time Artemis fell in love with Orion, the most handsome of men and cleverest of hunters. Apollo, her brother, was jealous and sent an enormous scorpion to attack Orion, who, although he fought valiantly, was no match for the scorpion, and escaped into the sea.

At that moment Artemis arrived, and Apollo persuaded her that the distant figure was a man who had insulted one of her priestesses. So Artemis took aim with her bow and arrow, and shot the figure dead.

Realizing her dreadful mistake, that she had killed the man she loved, she turned Orion into a constellation eternally pursued by a scorpion, which remains to remind everyone of Apollo's jealousy and lies. This myth could explain why the Moon's rays are said to be the arrows of Artemis.

But this chaste, virginal Artemis is only one aspect of the goddess. In other parts of Greece, such as Ephesus, she was an orgiastic nymph; at Patrae she was worshiped as the Lady of Wild Things; at Messene burnt sacrifices were made to her; and at Hierapolis the sacrifices were hung on the trees of an artificial forest inside her temple.

The most famous temple to Artemis (though others have claimed it as Diana's) was at Ephesus. Here, she was shown wearing a necklace of acorns, revealing her early connection with the woodlands, and a tower-shaped crown like the great ancient mother goddess, Cybele.

Artemis was the goddess of childbirth in this form – rather different from the virgin goddess! The statue of Artemis at Ephesus had an inscription reading "Aski. Kataski. Haix. Tetrax. Damnameneus. Aision." This probably meant something like: "Darkness-Light-Himself-the-Sun-Truth." Somewhat cryptic! Copies of the Ephesian letters were carried by people as a potent good-luck charm, and used by ancient magicians to cast out evil spirits.

The name Artemis probably originally meant "High Source of Water", reflecting how the Moon governs the waters. And sure enough, she was

Above: Diana, chaste goddess of the hunt.

Left: The Moon Goddess had many different faces. Her Medusa-like aspect is the fearsome Moon appearing on a storm-tossed night.

From graceful dancer reflecting the Moon's ethereal beauty, to strident huntress of the night: the Moon's qualities have been ascribed to woman herself, creating many forms of the Moon Goddess.

believed to rule the ebb and flow of both the physical tides, psychic powers and women's monthly phases. We may imagine the idea of Moon Goddesses to be long past, but the "Goddess Movement" is in fact getting stronger as people seem to be moving towards the ancient wisdom of, 'all things being connected', as fast as, on the other side of the scale, the industrial society is reaching its peak in destroying the very planet we live on.

Artemis was also worshiped in Sparta. The story goes that two young princes entered a thicket of willows (trees sacred to the Moon) and found a carved wooden image of the goddess. The sight of it so terrified them that they went mad. From then on, Spartan boys would vie with each other every year to see who could bear the most scourging before the statue. This is probably connected with the old idea that scourging was a means of purification, and whipping was once used to treat "lunatics".

Human sacrifices may also have been made to Artemis at Sparta. It is said that a quarrel arose between rival devotees who were sacrificing together at the altar. Many of them were killed in the sanctuary, the rest died of plague soon after. An oracle told them that they only way to restore peace and placate Artemis was by drenching the altar with human blood, and so they cast lots for a victim to be sacrificed. The ceremony continued annually until King Lycurgus replaced it with the altogether more civilized ritual of flogging boys until they bled. But even then, the carved image had acquired such a taste for blood that the floggers were urged by the priestesses to ever greater brutalities.

This, then, was the "dark form" of the goddess. Greek mythology depicts the goddess in three different forms, as in the three phases of the Moon: the young virginal girl of the waxing moon; the fertile mother of the full moon, and the terrifying old crone of the waning moon.

Helen and Helle are further local variants of the Moon Goddess.

Aphrodite, in her "orgiastic" mode, was also associated with the Moon; her priestesses despised the patriarchal system in which women belonged to their fathers and husbands. Nemesis, too, was a Nymph moon goddess, associated with a love chase which fits well with the Moon.

She was said to have been pursued by Zeus, though an earlier myth says that she chased him, and that he tried to escape by changing form but she did the same until she devoured him at the summer solstice. Hints, perhaps, of why the Moon keeps disappearing.

The face that launched a thousand ships . . . Helen of Troy, whose name is one of the many variants of the Goddess.

The Roman Artemis – Diana

Diana is the Roman counterpart of Artemis, being goddess of hunting and the Moon. She is the daughter of Jupiter. Some say that her name comes from the word *dies*, a day, others that it is from the Indo-European root di, meaning bright or shining. Coincidentally, the Celtic words *dianna* and *diona* also mean divine or brilliant. Perhaps the confusion just stems from the fact that worship of Diana was so widespread for such a long time. What is not in doubt is that she represented the Moon, just as Dianus, the god of light, represented the Sun. In a way she is a universal goddess, standing for many things but above all associated with the Moon.

Diana, like Artemis, decided to remain celibate, because of the pains she saw her mother suffer in labor. She also looked after other women in childbirth and dedicated herself to hunting, with her likewise celibate nymphs. She was probably originally a goddess of woodlands.

There was a famous Diana cult at Arica, where her shrine stood in a grove and she was worshipped along with a male god of the forest. The custom here was that the priesthood of the shrine was given to a runaway slave after he had plucked a branch from a certain tree in the grove and killed the previous priest in single combat. But as Diana became increasingly associated with Artemis, she gradually took on more characteristics of the Moon.

The temple of Diana (and/or Artemis!) at Ephesus was one of the seven wonders of the world. It was said that the statue to her had fallen from the heavens, and it was covered in many breasts, a potent image of fertility and motherhood. The cult of Diana became very important in Ancient Britain. She is supposed to have directed the Trojan Prince Brutus to take refuge in Britain after the fall of Troy. The London Stone, a sacred relic still preserved, is said to have been the first altar raised to Diana, and a temple to her also existed at Bath. The London temple was established by Brutus as a symbol of thanksgiving, possibly on the present site of St Paul's Cathedral.

The alchemists also revered figures of Diana, and regarded silver as her metal – the old custom of turning over silver at the New Moon is a relic of worship of the Moon Goddess. Such dedication to Diana has continued through the centuries, and has often been linked with the "old religion", and witchcraft. But more of that later! Let us end talk of Diana with a nineteenth-century fairy-story which is far removed from its original status, but which shows how, over the years, she became linked with anything that might possibly go bump in the night!

All things were made by Diana, the great spirit of the stars, men in their time and place, the giants which were of old, and the dwarfs who dwell in the rocks, and once a month worship her with cakes.

There was once a young man who was poor, without parents, yet he was good. One night he sat in a lonely place, yet it was very beautiful, and there he saw a thousand little fairies, shining white, dancing in the light of the full moon.

"Gladly would I be like you, O Fairies!" said the youth, "free from care, needing no food. But what are ye?"

"We are moon-rays, the children of Diana," replied one:

> *"We are children of the Moon,*
> *We are born of shining light;*
> *When the Moon shoots forth a ray,*
> *Then it takes a fairy's form."*

"We are children of the Moon.
We are born of shining light."

And thou art one of us because thou wert born when the Moon, our mother Diana, was full; yes, our brother, kin to us, belonging to our band."

The fairies went on to tell him that, because he is a full moon child, he has only to touch the money in his pocket at full moon, and say:

> *"Moon, Moon, beautiful Moon,*
> *Ever be my lovely Moon!"*
> *and his money would be doubled.*

But one month this did not work, and he asked the Moon for the reason.
A shining elf appeared, and told him that he also had to work for it:

> *"As appetite comes by eating and craving,*
> *Profit results from labor and saving."*[33]

And so a tale of the origins of the fairies ends up as a Victorian morality story!

The line of the Moon Goddess has thus continued, from the earliest forms of the divine woman, through goddesses of many different names.

Her power, influence and beauty have even appeared in the form of the Madonna, the Queen of Heaven, and her presence is to be found within our very unconscious. Moon, change, goddess, fertility, mystery – who can now untangle one from the other?

The Magic Cycles of the Moon

"Surely she knows
If she is true to herself, the Moon is nothing
But a circumambulating aphrodisiac
Divinely subsidized to provoke the world
Into a rising birthrate."[34]

Goddess Moon, protector of childbirth, regulator of the tides of Earth, the flow of life. An invisible umbilical cord seems to connect us with our nearest neighbor. Throughout the ages we have believed that the Moon affects a woman's fertility and can even determine the time of birth and death. Births are said to happen mainly around Full Moon, and deaths mostly when the Moon is waning.

We have known since ancient times that the life-cycles of plants, animals and the waters are intimately tied to the Moon. It is not surprising that one of the greatest mysteries – fertility and motherhood – should have been linked with it. After all, does the Moon not give birth to itself every month?

The Lunar "Period"

The Moon, it is said, governs a woman's menstrual cycle. Menstruation actually means monthly – occurring every lunar month. It is not a coincidence that the average time between menstruation is the time between two Full Moons. Not every woman in the world menstruates on the same day of the lunar month but the "period" between menstruation is rarely that of the calander month. Nature has little concern for man's inventions. It has been suggested that in ancient times, before there was any artificial light or any type of chemical contraception, women may have ovulated and menstruated throughout the world at about the same time, because of the Moon's influence. Modern communes have found that if women are living together in groups, they tend to menstruate all within a few days of one another, as though some hidden rhythm adjusts itself in harmony. In some societies women would go into isolation together during this time; and today in America some women use "Moon Huts" for the same purpose – not because they consider themselves unclean, but as an opportunity to celebrate the mysteries of their life forces together. The women will gather together at their "Moon Time" and connect with the "Great Mother" the earth, sitting and

allowing their own "Moon Time" discharges to seep back into the earth, giving to her from the source of their own fertility, and feeling the earth's pulse through their own bodies as a connection to the divine.

"The womb Rattles its pod, the moon
Discharges itself from the tree with
nowhere to go."[35]

Way back in ancient Greece, Aristotle wrote that menstruation begins when the Moon is waning. Darwin, commenting on the coincidence of the length of the lunar month with a woman's month, wrote: "Man is descended from fish . . . why should not the twenty-eight-day feminine cycle be a vestige of the past when life depended on the tides, and therefore the Moon?" If the belief is ancient, what is the modern evidence?

A Swiss Nobel Prize-winner, Svante Arrhenius, recorded 11,807 menstrual cycles. He found a tendency for the bleeding to begin more often during the waxing rather than the waning Moon, and particularly on the evening before the new moon. Other studies have both confirmed and refuted this. The similarity between the length of the female cycle and the lunar cycle has also caused much speculation. Some scientists have said it must just be coincidence, and have argued that women's cycles are not the same length as the lunar cycle anyway.

However, an extensive survey has shown that the average female cycle is 29.5 days, which is almost exactly the time of a lunar month.

Why should this be? One theory is that a substance called melatonin is produced during darkness, with a five-fold increase at night, and this inhibits hormone production. According to this theory, women's cycles have gradually evolved over the years, as women have imbibed moonlight. This is an attractive theory, and there may be something in it. But it does beg the question of why few other animals – who presumably are subject to the same lunar influence – have not developed the same rhythm. For instance, the rhythm for sheep is eleven days, for chimpanzees thirty-seven days – the opossum is one of the few animals to share the same cycle as us.

But such doubts have not been enough to deter everyone, and attempts have been made to base contraception on the lunar cycle. A Czeckoslovakian scientist, Eugen Jonas, confirmed that the time of ovulation is connected with the Moon, and also that this coincided with the phase the Moon was in when the woman was born (36). This led him to set up a system for working out a woman's contraceptive chart according to lunar phases. These were claimed to be ninety-eight per cent effective – although other obstetricians disagreed.

The Moon may have a lot to answer for! It has been called "the great midwife" for good reason.

Lunar Births

The Moon has also been linked with the time at which a woman is delivered of her child – it has been called "the great midwife". A 1950s study at the Tallahassee Memorial Hospital found a clear, statistically significant trend for more babies to be born within two days of the full moon than within two days of the first quarter. A New York City study looked at 510,000 births between 1948 and 1958. It also found a correlation with the Moon: the birth-rate was one per cent higher in the two weeks after the full moon than before. But on the other hand, other studies disagree with this – a 1973 study found the exact opposite, and a 1960s study found that births were actually centered on the time of the full moon. It seems that the elusive Moon is not to be so easily analyzed and dissected! Perhaps it is to the times of the tides, rather than the phases of the Moon, that we should look for an explanation of these mysteries. After all, the tides certainly have an effect on many other animals – and if, as Darwin pointed out, our fishy origins still have some deep-rooted effect on us, might not the tides sway us, too? Two independent German studies both found that there were, indeed, far more births at or just after high tide. Perhaps the connection is through gravity. Whatever the case, it is clear that, although man has set foot on the Moon, he has not yet uncovered all its secrets.

Lunar Sex

The Moon has also been said to determine the sex of a child. Eugen Jonas, he of the lunar contraception, found that he could predict a child's sex with considerable accuracy, according to the Moon's position in the sky at the time of conception. If intercourse was in a "male" sign of the zodiac (e.g., Aries), the child would be a boy, if in a female one (e.g., Taurus), it would be a girl. He apparently told 8,000 women who came to his clinic at what phase of the Moon they should conceive if they wanted a child of a particular sex, and ninety-five per cent were successful. According to Lyall Watson, the reason could be that the Moon produces changes in the Earth's magnetic field, which in turn influence the semen and sort it out into some kind of polarity.

Man in the Moon?

While the Moon may have its most dramatic or overt effect on women, we should not forget the men. They, too, have been shown to have some kind of "period" each month. A Japanese study found that bus and taxi drivers had particular times of the lunar month at which they were more susceptible to accidents. By rearranging each man's timetable in accordance with his "period", the accident rate dropped dramatically.

Generally, though, the Moon is most strongly associated with female fertility and motherhood. Does this mean that the Moon itself is considered to be feminine? Here is a paradox – it is seen as both genders. The Moon was worshiped as the "Great Mother" in various forms – Diana, Artemis, Ishtar of Babylonia, Asthoreth of the Phoenicians – but it has also been seen as capable of making women pregnant, and must therefore be male. The French and the Italian words for the Moon are feminine, the German is masculine. Some Eskimoes believe that the Moon is brother of the female Sun, while many Malaysians see it as the mother of the stars. There are many different versions. Maybe this is the nearest we can really come to giving the Moon a gender; that when the Moon takes some sort of action – such as the Eskimo Moon god who seduces women and makes them pregnant – then it displays the masculine qualities, but when the Moon must be approached, her receptivity invokes the female deity.

The evidence is mounting that the Moon does indeed affect the patterns of life. Who, on a moonlit midsummer's night, could doubt it for a moment?

Chapter 4

The Eastern Moon

"When a finger points at the Moon, the imbecile looks at the finger."
(Chinese proverb)

Imagine sitting on the seashore; it is a cloudy night, but as you gaze out across the dark waters the clouds break, and a full moon casts a glimmering path across the ocean. In that instant something deep within you reaches upwards and follows that shining path heavenwards, irresistibly drawn by the Moon just as the tides are.

The Moon has long been associated with man's spirit and his inner life. The Buddha, it is said, became enlightened under the Full Moon. In India, the Guru Purnima full moon is the time when everyone travels to be with their spiritual master. Each one of us somehow knows that the Moon links man with some greater truth.

Even in our rational western culture, there is the old belief that the Moon was the place where everything went that had been wasted on Earth. Misspent time, broken promises, unanswered prayers, unfulfilled wishes and intentions – all ended on the Moon. An old story tells of a man journeying there and discovering bribes hung on gold and silver hooks and wasted talents kept in vases, all neatly labeled.

> *"There broken vows and death-bed alms are found*
> *And lovers' hearts with ends of riband bound,*
> *The courtier's promises, and sick man's prayers,*
> *The smiles of harlots, and the tears of heirs."*[37]

From this old belief comes the phrase "The limbus of the Moon" – to be in limbo, waiting for the soul to be reborn. The very fact that there are so many expressions involving the Moon bears witness to the moon's strong effect on us all. It is in Buddhism that the Moon's affinity with our spiritual life is most clearly perceived, and finds its most beautiful expression. In Buddhism, the Moon usually symbolizes enlightenment.

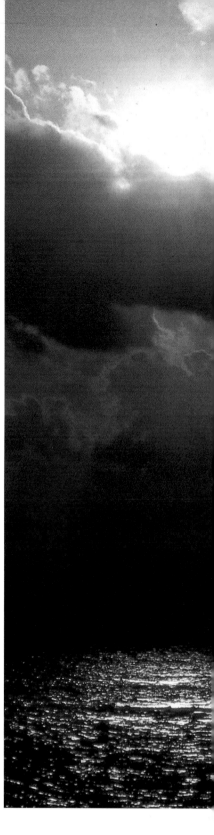

"Enlightenment is like the Moon reflected on the water. The Moon does not get wet, nor is the water broken. Although its light is wide and great, the Moon is reflected even in a puddle an inch wide. The whole Moon and the entire sky are reflected in dewdrops on the grass, or even in one drop of water.

Enlightenment does not divide you, just as the Moon does not break the water. You cannot hinder enlightenment, just as a drop of water does not hinder the Moon in the sky.

The depth of the drop is the height of the Moon. Each reflection, however long or short its duration, manifests the vastness of the dewdrop, and realizes the limitlessness of the moonlight in the sky."[38]

The words of the thirteenth-century Zen Master Dogen remain one of the most exquisite ways by which we may grasp something of the unknown: "Through one word, or seven words, or three times five, even if you thoroughly investigate myriad forms nothing can be depended upon. Night advances, the Moon glows and falls into the ocean.

The black dragon jewel you have been searching for is everywhere."

"Zen Master Guangzuo of Mount Zhimen was once asked by a monk, "What is going beyond buddha?"

He said, "To carry Sun and Moon on the end of a staff. This means that you are completely covered by the Sun and Moon on top of a staff. This is buddha going beyond. When you penetrate the staff that carries Sun and Moon, the entire universe is dark. This is buddha going beyond. It is not that the Sun and Moon are the staff. 'On the end of a staff' means the entire staff."

> "Contemplate on the 16th night koan.
> When body Moon tries for fullness, mind Moon starts to fade.
> If you have a clear idea of Moon, a Moon will be born.
> But how can mid-autumn Moon be grasped?"[39]

> "Cold lake, for thousands of yards, soaks up sky color.
> Evening quiet: a fish of brocade scales reaches bottom, then goes first this way, then that way; arrow notch splits.
> Endless water surface moonlight brilliant."[40]

Just as Buddha became enlightened when the Moon was full, so other seekers have felt its influence. The female mystic Chiyono became enlightened through the Moon. This is how it came about.

Chiyono wanted to become a nun, a sannyasin, but her great beauty got in the way. She was turned away by every monastery, out of fear that the monks would be distracted by her. In desperation she mutilated her face to make herself ugly. By the time she found a Master it was impossible to tell whether she was a woman or a man. She was accepted as a nun.

One day she was carrying a pail of water, and as she did so she watched the reflection of the Moon in its surface.

"Even reflections are beautiful, because they reflect the absolute beauty. A real seeker has known so much in the reflection, it was so beautiful, such music was there, that now a desire has arisen to know the source."

"Enlightenment does not divide you, Just as the Moon does not break the water.
You cannot hinder enlightenment, Just as a drop of water does not hinder the Moon in the sky."

And then the catastrophe happened. The pail suddenly split apart. The water rushed out, taking the Moon's reflection with it – and Chiyono became enlightened.

She explained what happened:

"This way and that way
I tried to keep the pail together,
hoping the weak bamboo
would never break.
Suddenly the bottom fell out.
No more water,
no more moon in the water –
emptiness in my hand.

"Enlightenment is like an accident. But don't misunderstand me – I am not saying don't do anything for it. If you don't do anything for it, even the accident will not happen. It happens only to those who have been doing much

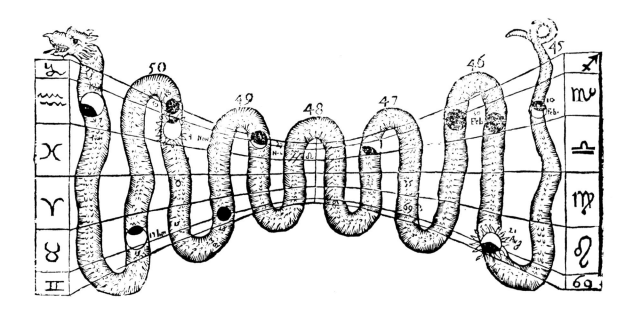

Top: A Chinese woodcut, originally used for predicting the Moon's eclipses.

for it, but it never happens because of their doing – and it never happens without their doing. All your meditations will just create an accident-proneness, an invitation, that's all . . . Without the invitation, the guest will never come."[41]

Another tale of the Moon and spiritual realization involves a Zen follower, Rengetusu, who was on a pilgrimage. As night approached she came to a village and asked for lodging for the night. The villagers must have found the sight of the old woman threatening, for they all refused her. She had to sleep out in the fields, on a freezing cold night, with only a cherry tree for shelter.

Awakened at midnight by the cold, she looked up. The cherry tree was in full blossom, with a misty Moon shining its radiance down upon her through the branches. Rengetsu was overcome by the sight and silently thanked the village for turning her away.

> *"Through their kindness in refusing me lodging I found myself beneath the blossom on the night of this misty moon."*
>
> *"A man becomes a Buddha the moment he accepts all that life brings with gratitude."*[40]

Even in the twentieth century, great teachers have taught that the Moon has a unique effect on our innermost lives.

Gurdjieff says that the Moon is not dead, but is a "planet in birth".

Living things feed the Earth, but everything that dies feeds the Moon.

You could say that the Moon is "hungry" and is fed by the soul at death. Gurdjieff talks of the "Ray of Creation." Life on Earth enables the Ray to grow, but it is the Moon that is the growing point of the Ray. One day, he says, the Moon will become like the Earth, and the Earth like the Sun. Then another Moon will appear, and so growth will continue.

Not only that, but the Moon has a very strong influence on our lives, far stronger than that of the Sun, he says. His "pupil", P.D. Ouspensky, explains:

"The Moon controls all our movements. If I move my arm, it is the Moon that does it, because without the influence of the Moon it cannot happen. The Moon is like the weight of an old-fashioned clock and organic life is like the clock mechanism which is kept going by this weight . . . it (the Moon) receives higher energies which little by little make it alive . . . like a huge electro-magnet, attracting the matter of the soul."[43]

So according to Gurdjieff the Moon has a "mechanical" influence on us, in effect making us act unconsciously and without awareness:

"We are like marionettes moved by wires, but we can be more free of the Moon or less free . . . by not identifying, not considering, struggling with negative emotions, and so on . . . All sleeping people are under the influence of the Moon. They have no resistance, but if man develops, he can gradually cut some of the wires that are undesirable and can open himself to higher influences . . . but nothing can cut him off from them except himself."[44]

The Moon, according to Gurdjieff, is a permanent center of gravity which balances our physical lives. It is necessary because we are so unaware, we do not normally have that balance. But once we do find it, we are no longer dependent on the Moon. As Master Dogen puts it:

"The Moon
abiding in the midst of serene mind;
billions break into light."

Above: A Chinese Moon Goddess holds a hare which was believed to live on the Moon.

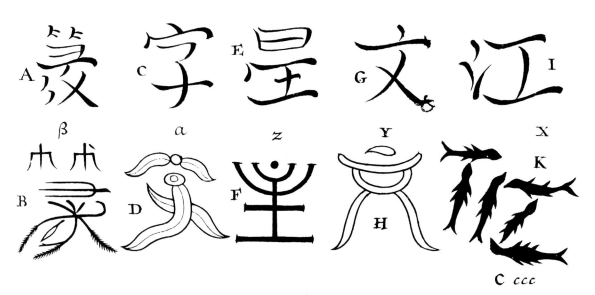

Chapter 5

Moon Words

Here are some of the many ways that the Moon's mysteries creep into our language, and thus into our everyday lives.

To shoot for the moon: to be very ambitious.

Over the moon: delighted about something – maybe shooting for the Moon proved successful!

Crying for the moon: longing for what is beyond reach, as children "cry for the Moon to play with". The French have a similar expression – *Il veut prendre la lune avec les dents*, meaning he wants to take the Moon between his teeth, from the old story about the Moon being made of green cheese.

Moonshine: illicit liquor, made by a "Moonshiner" in secret stills.

It's all Moonshine: it's nonsense, imagination, caused by the effects of the Moon on the mind. "O vain petitioner! beg a great matter; Thou now request'st but moonshine in the water."[45]

"I talk of things impossible and cast beyond the Moon."

"I care little about that nonsense – it's a' moonshine in water – waste threads and thrums, as we say."[46]

The Moon is made of green cheese: This was in common use by the early sixteenth century. "Green" refers not to the Moon's color, but to a new, immature cheese, round and uncut like the Moon, and whose mottled surface and color would also resemble the Moon.

To believe that the Moon is made of green cheese: to believe in the absurd: "You may as soon persuade some Country Peasants, that the Moon is made of Green Cheese (as we say) as that 'tis bigger than his Cart-wheel."[47]

Moonlight flit: undetected escape.

The man in the Moon: said by some to be a man carrying a bundle of sticks collected on the Sabbath. Some say he also has a dog with him. Another version is that the man is actually Cain, with his dog and thorn-bush. The thorns symbolize the fall, and the dog represents the foul, animal side of man. He has also been said to be Endymion, taken to the Moon by Diana.

Casting beyond the Moon: to make wild speculations – "I talk of things impossible, and cast beyond the Moon." (Heywood).

Moonrakers: a bunch of Wiltshire country folk were caught raking a pond in the middle of the night. When asked by the excise men what they were doing, they explained that they were trying to rake out the Moon. Hence the phrase has also come to mean simpletons.

Moon-drop: a substance which, in Roman times and later, was supposed to be shed by the Moon on herbs when an incantation was made:

"Upon the corner of the moon
There hangs a vaporous drop profound;
I'll catch it ere it comes to ground."[48]

Moonlighting: different meanings in various countries. In the USA it means having a night job as well as day-time employment; in Australia it describes riding after cattle by night; in the UK it means illicit work; and in Ireland it was violence carried out at night.

"You may as soon persuade some Country Peasants that the Moon is made of Green Cheese as that 'tis bigger than his Cart-wheel."

For Moonshine in the water – for nothing.

I know as much about it as the Man in the Moon: I know nothing! To moon over something or someone: somewhere between moan and swoon.

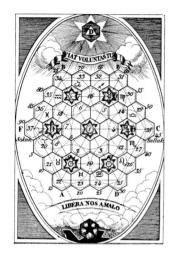

"To the moon, Alice!" – famous threat used by Ralph Cramden (Jackie Gleason) to his wife Alice.

Moon-calf: a name given to a repulsive abortion, which was once believed to have occurred due to the Moon's influence. It also came to be a term for an idiot or moron.

Minions of the Moon: night-time thieves. Also known as "Moon's men", and particularly referring to highwaymen: "The fortune of us that are Moon's – men doth ebb and flow like the sea."[49]

To find an elephant in the Moon: something that seems like a great discovery, but is just pure moonshine! The phrase came about when a seventeenth century man proclaimed with much pride that he had discovered an elephant on the Moon. It turned out that a mouse had crept into his telescope, and he had mistaken it for an elephant.

Diana's Worshipers: a name given to midnight revelers. They come home by moonlight, and so put themselves under her protection.

Moon about: wander listlessly, especially if in love.

Mooning: a rather more contemporary phrase, referring to a bare, moon-like posterior being displayed to passers-by.

Once in a blue Moon: very, very, rarely!

Moon-blindness: colloquial term for night-blindness, also called mooneye.

Moon Child: someone born under the sign of Cancer.

Then come the associations with the word lunar . . .

Lunatic: crazy person.

Lunacy: craziness.

Loony: crazy.

The ever-changing Moon has added a richness and poetry to our very language.

Chapter 6

O Luna

What is there in thee, moon, that thou shouldst move
My heart so potently?[50]

Tales of spells cast at the time of the full moon, witches flying by its mercurial light, of incantations made to the Moon Goddess; sorcery, magic and witchcraft are associated in most people's minds with the Moon. Why should this be? In this chapter we take a look at the Moon's bewitching ways.

In more primitive – some might say more earthy – times, man and woman were more aware of the flow of nature than are the people of this technological age. And just as the seasons were more keenly felt, so everyone would have been aware of the Moon's phases. How many of us today know exactly when and where the Moon will rise? What this all comes down to is that the Moon was a major part of people's lives – it even foretold the coming of the "moon-blood" or a new baby. As we have seen, the Moon goddess, with all her powers, was a potent and honored force in times past. But let us not forget that this goddess sometimes had some pretty nasty, bloodthirsty qualities too. And it is from this same source that "drawing down the Moon" comes from. The expression – and the practice – can be traced right back to the ancient Greeks, who said that the witches of Tuessaly had the power to draw down the Moon.

By this, they meant that the Moon also had an evil aspect, which could be manifested on Earth by those with occult powers. The witches made use of the Moon's evil influence in their spells and rituals. In later times, it was said that magicians could make the Moon drip poison into a bowl of water, which would hiss and bubble and then be ready for the magician's rites.

Left and top: The Moon has its more horrifying aspects too. It is thought to bring out man's most devilish and hellish behaviour.

Drawing Down the Moon

"Drawing down the Moon" is not such a simple task. It has taken many different forms throughout our varied history but fundamentally it requires a formula which results in the powers of the Moon Goddess being brought to Earth. And this is strong stuff indeed!

The connection between witchcraft and the Moon suffered some serious relapses in medieval Europe in favor of the male Lucifer. Perhaps the strength of Christianity had something to do with this. The popular image of the witch silhouetted on her broom against the crescent moon seemed to be all that had

survived. But was this strictly true? It is interesting that the link has been very firmly restored by modern witches. Perhaps the canny Moon was just lying low for a while.

In Italy, as recently as a hundred years ago, witchcraft and the Moon goddess were still hand in hand. The Catholic Church had been trying for years to quash witchcraft, but the old religion was still widespread. The Church claimed that the witches were worshiping Satan, but it was probably the Moon goddess Diana who they revered. They were using semi-religious, semi-magical practices long associated with her.

And no wonder the Church found it disturbing! For the witches believed that creation stemmed not from the masculine but from the feminine principle. Lucifer, or Light, was just the heat hidden in the mysterious depths and darknesses of Diana, as heat is hidden in ice.

The traditional association of witches with the Moon Goddess has continued – but mainly as the old crone beneath the Moon.

Italian peasant women are known to have used a little tome called the *Vangelo delle Streghe*, or *Gospel of the Witches*. Here is how this revered book talks of our origins:

"Diana was the first created before all creation: in her were all things: out of herself, the first darkness, she divided herself; into darkness and light she was divided. Lucifer, her brother and son, herself and her other half, was the light".[51]

The *Vangelo* explains the long-standing association between witches and the Moon Goddess. When Diana saw the beauty of Lucifer's light, she was filled with a desire to receive light back again into her darkness. But Lucifer

Left: The androgynous figure of Baphomet sits between the dark and light Moon, good and evil, his horns echoing the crescent shape.

Above: The winged Lucifier largely usurped the Moon Goddess in witchcraft until recent times.

fled from her and Diana went in search of advice "to the fathers of the Beginning, to the mothers, the spirits who were but the first spirit" – what we today might call her unconscious, and what Jung would call Ouroboros, the male/female foundation of Nature.

Diana was told: "To rise she must fall; to become the chief of the goddesses she must become a mortal." And so, as Lucifer had done, she descended to Earth, where, according to the *Vangelo*, she carried out the first ever act of witchcraft. Her brother Lucifer had a beautiful cat which always slept on his bed at night. Diana knew that it was really a fairy spirit in disguise, and she persuaded it to change forms with her so that she might lie on her brother's bed. In the dark of the night she returned to her own form, and made love with her brother in his sleep. Through this she became pregnant, and eventually gave birth to a daughter, Aradia.

Lucifer, on awakening, was furious that "light had been conquered by darkness". But again Diana cast a spell over him, this time by singing to him until he was again calmed and enchanted: "It was the buzzing of the bees, a spinning-wheel spinning life. She spun the lives of all men; all things were spun from the wheel of Diana. Lucifer turned the wheel".

Then Diana, again by witchcraft, created the heavens with their stars, and made rain fall on the Earth. She became Queen of the Witches, "the cat who ruled the star-mice, the heaven and the rain". Diana's influence seems to have been thought of by the witches as benevolent but far from compliant! She saw the feudal rich and the Catholic Church oppressing the hungry poor, and sent her daughter Aradia to Earth to be the first witch and to help them. Just as Diana instructed her daughter in witchcraft, so Aradia later instructed her earthly followers. When Aradia left Earth she decreed that her witches should meet every month at the full moon in an isolated place, where they should adore the spirit of Diana, their Queen. In return, Aradia said, Diana would continue to teach them the secrets of witchcraft.

As a sign of their freedom to do as they wished, the witches were told, their feasting and dancing should be carried out naked: "They shall dance, sing, make music, and then love in the darkness, with all the lights extinguished for it is the Spirit of Diana who extinguishes them." And at their suppers they should eat cakes made from meal, wine, salt and honey, in the shape of a crescent Moon, and consecrated to Diana. This incantation was to be made before eating them:

"I do not bake the bread, nor with it the salt,
Nor do I cook the honey with the wine;
I bake the body and the blood and soul,
The soul of great Diana, that she shall
Know neither rest nor peace, and ever be
In cruel suffering till she will grant
What I request, what I do most desire,
I beg it of her from my very heart!
And if the grace be granted, O Diana!
In honor of thee I will hold this feast,
Feast and drain the goblet deep,
We will dance and wildly leap,
And if thou grant'st the grace which I require,
Then when the dance is wildest, all the lamps
Shall be extinguished and we'll freely love!"

So it seems that despite the Church, worship of Diana was widespread for many centuries, and its last surviving echoes may be what is left of witchcraft. She even became associated with those friends of the witches, the fairies. Titania, in Shakespeare's *A Midsummer Night's Dream*, bears another old name of the Moon Goddess, and the play makes many references to the "underworld" and the Moon:

"Oberon: Ill met by moonlight, proud Titania.
Titania: What, jealous Oberon! Fairies, skip hence;
I have forsworn his bed and company . . .
Therefore the moon, the governess of floods,
Pale in her anger, washes all the air,
That rheumatic diseases do abound . . .
And this same progeny of evils comes
From our debate, from our dissension:
We are their parents and original."

There are echoes in this of the old legends of quarrels between Sun and Moon, Diana and Lucifer. All is not right with the world while male and female principles are quarreling.

Right: Witches concoct an ointment used for flying on the Sabbath.

Who Knows the Hooked-Nose Horror?

What, though, of that old hook-nosed witch hag so beloved of fairy stories? Diana moves over, and in comes Hecate, symbol of the old-woman, waning phase of the Moon. This fearsome old creature was originally the ancient Greek goddess of witchcraft, but later came to be associated with the Moon and other Moon goddesses. The Romans depicted her enthroned in triple form, with three hands and three pairs of arms, holding daggers, whips and torches, and with serpents at her feet. She was also shown with howling dogs, probably because dogs howl at the Moon, and she is often invoked by magicians and witches for her underworld connections. The sign used to invoke her is a crescent moon with the two points upwards, and a third point in the middle between them. She is goddess of darkness, the dead, ghosts and terror, and was believed to frequent tombs and drink the blood of corpses, and to inflict madness and epilepsy on the living. Medieval witches would set out offerings of dogs' flesh for her at crossroads: she was the Moon in its spine-chilling aspect:

"You who rejoice to hear the barking of the dogs and to see the blood flow; you who wander among the tombs in the hours of darkness, thirsty for blood, and the terror of mortal men . . ."

Lovely Lilith

Lilith was another aspect of the Moon, this time the archetypal seductress, the *femme fatale*. And rumor has it that the witches were not loath to draw down this part of the Moon if the need arose! Lilith was another patroness of witches, the beautiful vampire who would suck a man's life-blood as soon as look at him! However, her divine beauty did have one flaw – she had clawed feet like a bird of prey. In medieval France she was known as la Reine Pedauque, the queen with a bird's foot, and was depicted flying by night as the leader of phantoms who were said to live on the Moon.

This form of the Moon goddess personifies all the erotic dreams that haunt man with their forbidden delights. The Jewish people even made amulets to protect themselves against her, and their legends say that she was Adam's wife before Eve, but that she only came to him and made love with him in his dreams. Thus she, too, became associated with the unseen beings of the world, the fairy races.

The Anglo-Saxons also had a tripartite goddess who became linked with witchy goings-on. She was called Wyrd (the origin of the word weird), and like the Moon was thought of variously as a young woman, a mature one, and an old crone.

"Our Lady of the Moon, enchantment's queen,
And of midnight the potent sorceress,
O goddess from the darkest deep of time,
Diana, Isis, Tanith, Artemis,
Your power we invoke to aid us here!"[52]

Left: The Moon hag whose look may chill the blood is a familiar image of witchcraft.

Below: Lilith, one of the three aspects of the Moon Goddess in witchcraft, is the archetypal seductress.

Illusions of Baphomet

Baphomet is a mysterious figure, often associated with witchcraft. It is both male and female, showing characteristics of both, and usually depicted with the Moon. It is often used as the figure for the devil in tarot cards but was probably originally the god Pan, symbolizing the whole of nature – hence the Moon. Again there is a connection with Diana, in the witches' tales of how Diana seduced Lucifer.

Baphomet is shown with curved horns, as devils often are. This is probably another Moon allusion. They may be seen as representing either the crescent moon, or the two horns of the Moon symbolizing both good and evil. At one time people believed that the Moon was the Devil's home, or at least that

devils occupied the space between the Earth and the Moon. Thus, just as the Virgin Mary is sometimes shown standing on a crescent Moon, as though having conquered it, so the crown of horns on the Devil may show that he is still ruled by the evil realm.

Making the Moon Magic

It is not surprising, with all these Moon goddesses and symbols around, that witchcraft has many rites and rituals involving the Moon. One of the best known of these is scrying, or crystal-gazing, which has long been used by witches to develop their clairvoyance. The ease with which it happens varies according to the phases of the Moon. But there is no one phase that is universally better than another – everyone has to find their own.

Most people think of scrying as staring into a crystal ball. But many other objects have been used by witches. That old witchy prop the cauldron is one such; its dark interior lends itself well to the art – when filled with water and with a silver coin placed at the bottom of it, it resembles the Moon shining in the dark of night. The witch or seer gazes softly into the waters until an image comes, either directly to the eyes or in the mind, and usually in response to a question about something.

A magic mirror has the same qualities as a crystal ball, but must be made in time with the Moon. Witches say that it is particularly effective to make and consecrate your own. A round, concave piece of glass is needed, such as that from an old picture-frame or clock-face.

When the Moon is waxing, several coats of black enamel paint are applied to the back of the glass (that is, the convex side). It must never come into contact with direct sunlight, for that would deplete its sensitivity, but moonlight will charge it up. When the Moon is full the mirror is consecrated before it is used, often with this incantation:

"Round of silver shining bright,
As the moon at still midnight,
When the witching hour has struck,
Shadows show of life and luck.
By this rune be now enchanted,
And the second sight be granted."

Never use the mirror for anything other than magic, and expose it to moonlight at least three times a year.

The mirror may also be used for past-life discoveries. Set a white candle in a darkened room so that it lights your face but is not reflected in the mirror. Stare into your reflection, and say "Oracle of lunar light, send me now the second sight". As you gaze into your own eyes, as softly but unblinkingly as possible,

"Oracle of lunar light, send me now your second sight."

Left: The joyous rites of Pan, Moon-horned god of nature, are part of the pagan inheritance of witchcraft.

your image will subtly change, to something with which you are, somehow, strangely familiar.

Another variation of scrying uses the Moon more directly. When the Moon is full, at its peak in a clear sky, take a small round convex mirror out of doors. Find a comfortable position by yourself, where you can catch the Moon's reflections in the mirror. Gaze at the pinpoint of light, and gently tilt the mirror by tiny, tiny amounts, so that the Moon's image shifts and moves. You can then explore the Moon's effect on your psychic being.

Try, too, watching the reflections of the Full Moon on the sea. Just sit on the beach and allow your eyes to follow the path of the moonlight to the horizon and back, repeating this until you wish to close your eyes. Then take inside you whatever image is left upon your eyes, and feel that gentle light kindled in you to the sounds of the ocean.

Lakes, sometimes called Diana's mirrors, may be used for scrying in the same way as a crystal ball or mirror. On the night of the full moon, lie down beside the lake and stare softly at the reflection of the Moon on the still, black surface of the water, asking Diana for help, until messages and symbols emerge. If you have a particular question, then ask it, otherwise just allow through whatever life wants to tell you.

"For she, too, loved the solitude of the woods and lonely hills, and sailing overhead on clear nights in the likeness of the silver Moon looked down with pleasure on her own fair image reflected on the calm, the burnished surface of the lake, Diana' mirror."[53]

The Moon's reflection in water can be used as a potent means for exploring your inner being . . . and maybe discovering a few lunar secrets.

Lunar Secrets

The lunar/witch connections are innumerable. The various numbers which we associate with witchcraft also have lunar origins. The number seven, for instance, is held in reverence. It comes from the "Sacred Seven", the seven heavenly bodies of Saturn, Jupiter, Mars, Sun, Venus, Mercury and Moon. Witches probably inherited this from the ancient astrologers, who believed that all things were ruled by them.

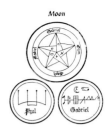

Moon

The number thirteen is also associated with witches – in popular imagination it is thought to be the number in a coven – but originates with the thirteen lunar months. Similarly the number three (remember how a spell is always made three times?) comes from the phases of the Moon. The ancient Druids said there were: "Three embellishing names of the Moon: the Sun of the Night, the Light of the Beautiful , and the Lamp of the Fairies."

Witches' amulets often depict a Moon. A certain Italian one, for instance, is in the form of a sprig of rue and vervain, said to be the two plants most loved by Diana. It is, of course, made of silver, Diana's metal, and contains a waning Moon to protect the wearer from evil.

Beware of crossroads in the dark of the night! They are the traditional meeting-place for witches, because of the witches' tripartite imagery of the Moon goddess. Statues of Diana or Hecate were set up by the Greeks and Romans where three or more roads met, and so crossroads became sacred to the Moon goddess. Remnants of the practice may still be found today in England. In Ashdown Forest, Sussex, Wych Cross, where three roads meet, was originally called "Witch Cross" because it was a meeting-place for witches. And in the New Forest, Hampshire, witches met at a forest crossroads called Wilverley Post, near an old oak tree called the Naked Man.

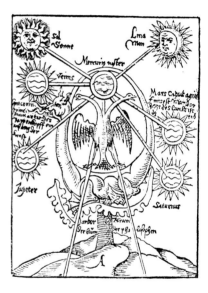

As for that most notorious of witch occupations, the casting of spells, the Moon is of utmost importance. Spells are traditionally cast according to Moon phases, the waxing moon being the time for benevolent magic, the waning for magic with more sinister intentions.

> *"Pray to the Moon when she is round.*
> *Luck with you will then abound.*
> *What you seek for shall be found,*
> *On the sea or solid ground."*

Lunar Rituals and Lunar Spells

Although there are many ancient spells, you can create your own powerful magic by inventing rituals that fit with the Moon. Two contemporary American "witches" tell of their methods:

"Last night we hung out of the east windows and howled at the Moon, incredible orb gliding up over the eastern hills . . . and made up a song to her. During the night I fell into a dream that enabled me to go under the belly of death, as the giver of life . . ."[54]

And in celebration of the New Moon:

"We sit in a large circle in front of the cabin. We join hands and follow each other down to the meadow, down into darkness. We tell stories of darkness . . . There are images of fear as well as power and strength expressed, a lot of images of calm, warmth and rest. A large candle is lit . . . (Billie) has made 10 small bags with drawstrings, each has a black bead attached to the drawstring, signifying the dark Moon. She gives them to us to keep . . . We find seeds inside the bags. Seeds, the small beginnings, the New Moon . . ." (Adler, op.cit.)

Mooning for Love

If you wish for a little help with your lunar spells, here are a few to try. This first one, though, is put in for its entertainment value rather than its moral virtue! It comes from *Aradia, or the Gospel of the Witches*,[55] a nineteenth century book.

"When a wizard, a worshiper of Diana, one who worships the Moon, desires the love of a woman, he can change her into the form of a dog she, forgetting who she is . . . will at once come to his house, and . . . take on her natural form and remain with him. And when it is time for her to depart, she will again become a dog and go home, where she will turn into a girl. And she will remember nothing of what has taken place . . . And she will take the form of a dog because Diana has ever a dog by her side.

"This is the spell to be repeated by him who would bring his love to his home:

"Diana, beautiful Diana!
Who art indeed as good as beautiful,
By all the worship I have given thee,
And all the joy of love which thous hast known,
I do implore thee aid me in my love!"

Drinking the Moon

Another spell from the same book calls on Diana to help with the wine-making.

"He who would have a good vintage and fine wine should take a horn full of wine and with this go into the vineyards . . . and then drinking from the horn, say:

"If drinking from this horn I drink the blood –
The blood of great Diana – by her aid –
If I do kiss my hand to the New Moon,
Praying the Queen that she will guard my grapes . . .
So may good fortune come into my vines."

Lunar Love-Charms

This invocation is probably of great antiquity, and it is interesting that it involves a horn, the symbol of the new moon, sacred to Diana. (It is said that Apollo built an altar to Diana, made entirely of horns.)

Love charms often make use of lunar phases; for instance, to see a future lover in a dream, gather yarrow when the new moon is visible.

Then place the flowers under your pillow, and repeat this verse:

> *"Thou pretty herb of Venus' tree,*
> *Thy true name it is yarrow;*
> *Now who may bosom friend may be*
> *Pray tell thou me tomorrow."*

It is to be hoped that the results are better than the rhyme!

Mooning for Money

This spell is for receiving money. Set a dish, ideally of silver, containing water in the light of a waxing Moon, so that the Moon is reflected in it. Dip your hands in the water, and as they dry naturally, imagine money coming to you. It should arrive from an unexpected source before the Moon again reaches the same point in its cycle.

Spells and incantations, incense and healing plants . . . all are used by witches to achieve their ends, but must be used according to the Moon's phases.

Moon Diets

This next spell must surely have been devised by modern witches, as it calls on the Moon to help with dieting, a very twentieth-century concern. In pencil, draw as perfectly as possible an outline of the shape you wish to be. Around this perfect image draw the shape you actually are now. Be honest! Put the drawing in a safe place until the time of the full moon. Then smooth down a tiny part of the larger figure, symbolically removing weight from your body. Repeat this every day for fourteen days while the Moon wanes. Since results are likely to be gradual, repeat the process at the next full moon. (Scott Cunningham, *Earth Power* Llewellyn Publications, Minnesota, 1986)

Tugging at the Moon

A spell to attract love to you: for this you will need dried rose petals, a pinch of catnip, half a handful of yarrow, and a pinch each of mint, coltsfoot, strawberry leaves, ground orris root, tansy and vervain. These should be mixed together on a Friday evening when the Moon is waxing, and divided into three.

Take one of the parts out of doors, naked. Go down on one bended knee and throw the mixture up to the Moon, asking that love be sent to you.

Back inside, scatter another third around your bedroom. Sew the last portion up in a green or pink cloth, and wear this on your body. The smell alone should be enough to attract love to you![57] [58] Incense is traditionally used in incantations to the Moon, to add another element to the vibrations. Making your own adds another depth to that, through the energy you direct into the manufacture. The mixed incense can be burned over charcoals in a censer, or scattered on a fire.

Blessings

This recipe is specifically for use during the full moon, to receive its blessings and to use in any ritual associated with the Moon. Mix together equal parts of ground frankincense and white sandalwood, add a quarter-part of orris root and a few drops of lotus oil, and burn at the full moon.

This incense calls on all the planets, including the Moon, as it contains ingredients sacred to them all, and it is powerful when used during rituals. Mix together, in equal parts, frankincense (Sun), orris root (Moon), lavender (Mercury), rose petals (Venus), dragon's blood (Mars), cinquefoil (Jupiter), Solomon's seal (Saturn).

And here are two incenses which depend on the Moon's phases for their preparation. The first is for prosperity: mix equal parts of ground cloves, nutmeg, lemon balm, poppy seed and cedar, then moisten with a few drops of honeysuckle and almond oil. Do this on a Thursday when the Moon is

waxing. The second, a love incense, should be made on a Friday during the waxing Moon. Mix equal parts of ground rose petals, cinnamon, patchouli and red sandalwood, and burn during rituals for attracting love.

Chasing Love Away

If you would rather ward off love – or, at least, someone's unwanted attention – use camphor gum. Camphor is said to be ruled by the Moon. If you can persuade the undesired person to smell it, it will turn them right off you! Camphor can also be used in incenses to help sleep, and can be worn in a pouch around the neck to ward off colds.

Cucumber is also said to be a plant of the Moon. It supposedly aids fertility, if kept in the bedroom! Other plants said to be governed by the Moon and which can be used in rituals are: gardenia, to attract the opposite sex and as a link with the Moon; lettuce – the juice rubbed on the forehead will induce sleep, and the leaves when eaten will cool down lust; poppy – the seeds added to food will help make a woman pregnant. To affect your dreams, take a dried poppy seed-head, cut a small hole and remove the seeds, then write a question on a piece of paper which is then stuffed inside the pod and put beside your bed – your dreams should provide the answer to your question; sandalwood, also governed by the Moon, is said to be a good room purifier and of use in healing oils and incenses; finally, willow has all sorts of witchy connections, and is very strongly associated with the Moon, probably because of its watery origins. A willow wand is involved in healing rituals, and willow is used to bind a witch's broom. Willow can bring down the Moon's blessings, and planted near the house will protect the home. Carrying a piece of willow will even quell your fears of death.

The Final Draught

And, lastly, what if we wish to "draw down the Moon?", to follow the ancient practice? There are many different versions; ignore those which involve a cup and dagger for they are really solar rituals. This "recipe" for drawing down the Moon has all the main elements needed.

This ancient rite is performed out of doors, in the first three nights of a New Moon, soon after sunset, and needs three people, of whom at least two should be women. A bowl, preferably of silver or glass, is set between the participants, along with a small round mirror and a bottle of white wine or spring water. One woman holds the bowl, another pours wine into it, the third holds the mirror in such a way that it reflects the Moon into the bowl. Once the image has been directed there, one of the Moon goddesses – such as Diana, Artemis or Isis – is called upon to bring her blessings and powers into the wine or water. Prayers, incantations and chantings may be offered in return for the Moon's wisdom.

Next, thanks are given and a few drops of the precious wine are poured into the earth, whence it originally came. The wine is then drunk, as a communion with the Moon goddess. If water has taken the place of wine, it may be used for blessing and scrying, but should be discarded once the Moon is full. Special moon biscuits (see Part 2 for recipe) may be eaten with the wine, and a few crumbs scattered in thanks. This simple but potent ceremony, if done with love and respect, can bring us back in touch with our inner powers, our connection with the Moon goddess.

Alchemy

An important early sixteenth-century alchemical text, *The Rose Garden of the Philosophers*, depicts a king and queen standing on a Sun and Moon, with crossed flowers and dove above. The words beneath the picture read: "Mark well: in the art of our magisterium nothing is concealed by the philosophers

except the secret of the art, which may not be revealed to all and sundry. For were that to happen, that man would be accused: he would incur the wrath of God and perish of apoplexy. Wherefore all error in the art arises, namely, because men do not begin with the proper substance."[59]

Thus are we warned not to tamper with the alchemical arts. In this connection it may be noted that the left hands of the figures, not the right, are joined, – that is, alchemy is the way of intuition and creativity. This is the coming together of opposites, man and woman, Sun and Moon. For as the text also says: "For this work you should employ venerable Nature, because from her and through her is our art born and in naught else: and so our magisterium is the work of Nature and not of the worker."

The coming together of opposites – one of which is clearly represented by the Moon – is seen in the crossing of the stems of the flowers, one carried by a

Left: The Solar King and Lunar Queen of the medieval alchemists symbolize the joining of opposites.

dove descending from a star. They join east and west, south and north, above and below, Sun and Moon, man and woman. These opposites, say the alchemists, must be brought together to make one whole. Thus the solar king and lunar queen are symbols of a process which was supposed to be occurring in the alchemists' own experiments.

These are down to earth and practical, but attempting to bring in the subtleties of life's mysteries and unknown forces they hoped to be truly creating something new.

The text continues: "The meeting of the eyes (of the king and queen) having communicated its import to the noble, ready hearts . . . the left hands – of the heart side – reached spontaneously forward, and the right – of the spirit – crossed flowery tokens of the shared ideal to be realized: not the common, of desire, but the noble, of the self-loss of each in the identity that was theirs beyond space, beyond time."

Left: Solar King and Lunar Queen submerge in the mercurial liquid of consummation.

Below and right: The Moon is consistently found as a symbol of the anima throughout alchemical pictures.

The next picture, of the solar king and lunar queen in the mercurial bath (although the Sun and Moon are no longer actually depicted), shows that a merging of opposites has taken place. The flowers now form a circular movement, and all are joined together in the mercurial waters.

Then comes a picture of coitus between the king and queen, disturbingly suggesting a descent into chaos, perhaps the primordial forces of creation itself. They are submerged beneath the waters, but in the process are creating a new soul. The rhyme beneath the image reads:

"O Luna, folded in my sweet embrace
Be you as strong as I, as fair of face.
O Sol, brightest of all lights known to men
And yet you need me, as the cock the hen."

The Rosarium says of this picture: "Then Beya (the maternal sea) rose up over Gabricus and enclosed him in her womb, so that nothing more of him was to be seen. And she embraced Gabricus with so much love that she absorbed him completely into her own nature, and dissolved him into atoms."

No wonder man has feared the Moon!

Carl Jung cast fresh light on the Sol and Luna king and queen in the 1920s. He put forward the idea that they are symbols of the conscious and unconscious. It could be said that the ego-realm of the Sun king was investigated by Freud and his followers, while Jung (interestingly originally a follower of Freud too), plumbed the depths of the Luna Queen. Jung named the two forces *animus* and *anima*; the unconscious tends to personify itself in a female form in the man, but assumes a masculine form in woman. The *anima* is made up of many ancient images of woman, she is the "eternal woman" – and is often symbolized by the Moon.

The alchemists used to make an amalgam of crystallized silver, made from mercury in a solution of silver. It was known as Diana's Tree, or the Philosopher's Tree, because for the alchemists silver was Diana's color.

In alchemy each planet had an associated metal, the names of the two being virtually interchangeable. They were supposed to have a direct influence on a person's character and health. So somebody who was moody, for instance, would be considered a Saturn type, and would be treated with silver, through which he would be taking in the Moon, thus making him more quick-moving to dispel his gloom.

Chapter 7

Moon Madness

"It is the very error of the moon:
She comes more nearer earth than she was wont
And makes men mad. "[60]

The Moon has been blamed for many of man's misdemeanors. We are all, it is said, inclined to be a little more crazy – or "lunatic" – at full moon. The Moon even drives otherwise upright citizens to sprout hair, crawl on all fours – and become werewolves. At the very least, it is thought that we have more psychic powers around the time of the full moon.

In this chapter we look here at tales of all kinds of Moon madness, and ponder on the truth behind the stories. Does the Moon really make us all a little loony?

Where are the Werewolves?

"They had come to a point on the thoroughfare where it was bordered with the tombs of the dead, as the roads beyond the gates of any Roman city were likely to be. The light of the moon was almost as bright as the midday sun, when his mate stepped aside among the monuments . . . he stripped himself and put all his garments at the side of the road . . . he stood like a dead man. The soldier urinated so as to circle his clothes completely with water and then suddenly turned into a wolf . . . he began to howl like a wolf, and fled into the woods. Niceros at first scarcely knew what to do, but then went to pick up the man's clothes only to find that they had all turned to stone. Although he was in mortal alarm, he drew his sword and kept slashing shadows with it all the way until he reached his sweetheart's villa . . . she said "A wolf entered our place, worried all our sheep, and bled them like a butcher. But he didn't get the laugh on us even if he did escape; for our slave pierced his neck with a spear."

Niceros left for home at daybreak, and on his way passed the place where the garments had been turned to stone. Now he found nothing there but a pool of blood. When he reached the house . . . a doctor was attending to a wound in his neck. Then he realized that the man was a turn-skin, a werewolf."[61]

Such spinechilling accounts of werewolfism are found all over the world. There is something in it that grabs everyone's deepest fears. The belief can be traced as far back as ancient Greece; the Navaho Indians thought they rustled sheep and exhumed the dead; the Book of Daniel claims that King Nebuchadnezzar suffered a depressive illness which ended in him believing himself to be a wolf. In sixteenth-century Europe many men were put on trial for lycanthropy (form of madness in which patient imagines himself an animal and exhibits depraved appetites, change of voice, etc.), found guilty and executed.

The light of the silvery Moon has always been blamed for that most bizarre of human afflictions – lycanthropy.

Left: Even the Egyptian Nebuchadnezzar is believed to have suffered from a form of lycanthropy.

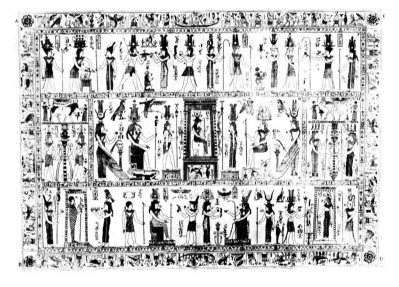

But whatever the country and whatever the circumstance, it has always been blamed on the full moon.

A nineteenth-century clergyman describes what happens when the Moon is full:

"The desire to run comes upon them. They leave their beds, jump out of a window, and plunge into a fountain. After the bath, they come out covered with dense fur, walking on all fours, and commence a raid over fields and meadows, through woods and villages, biting all beasts and human beings that come in their way. At the approach of dawn, they return to the spring, plunge into it, lose their furry skins, and regain their deserted beds."[61]

The stuff that nightmares are made of . . . the strange light that plays on us at night gives birth to hideous imaginings.

Spotting the Wolf

How do you recognize a werewolf. Here are some signs to watch out for, if your nearest and dearest start behaving a little strangely: a devil's mark on the buttocks; the presence of a bushy tail; eyebrows that meet between the eyes. This idea was promoted by the Danes, who presumably thought that the Mediterranean countries were full of werewolves; any wound inflicted on a werewolf remains after the beast has returned to human form; all werewolves, when not transformed during the full moon, have hair on the inside of the skin. A common test to identify a suspected werewolf was to peel back his skin to see if the hair was there! The discovery that there was no hair there usually came too late for the suspect.

Becoming a werewolf is evidently much easier than becoming a vampire.

For a start, you do not have to be dead. All that is needed is a full moon, perhaps with the aid of a curse, or some prop like a belt of human skin. After that, it could happen to anyone. The werewolf, unlike the vampire, is horribly alive. Black magicians could allegedly change themselves into werewolves at will, by rubbing a magic ointment into the naked body and putting on an enchanted belt made of wolf skin or human skin.

Lycanthropy

If this all seems like the stuff of fairy-tales, think again.

Lycanthropy has had a very real effect on people's lives. Confessions of werewolfism were quite common in medieval Europe. Perhaps this is not surprising, when we consider that red-hot pokers and the rack were used to get at "the truth". Execution usually followed a confession, generally by the normal means of hanging or burning, but occasionally, for that extra poetic touch, with a bullet made of silver, the Moon's metal.

Poor Peter Stump was put to death for this crime, in Germany in 1589.

A contemporary account tells of his fate: "The flesh of divers partes of his body was puled out with hot iron tongs, his armes thighes and legges broked on a wheel, and his body lastly burnt. He dyed with very great remorce, desyring that his body might not be spared from any torment, so his soul might be saved."

In 1521 three people suspected of lycanthropy were publicly executed.

In 1573 a local French parliament ordered citizens to try "with kitchen-spits, halberds, spears and sticks to hunt, capture, bind and kill the werewolf who infests the district". Perhaps this was all set off by the salacious tales told by Frenchman Jean Peyral, who was convicted of being a werewolf in 1518. The courtroom was bursting at the seams with people eager to hear his stories of how he copulated with female wolves and was turned into a wolf by the devil. Even the foul smell of the ointment he used in this process, and which he showed in court, was not enough to deter the "audience". Peyral was tortured before being sentenced, then burned and his ashes scattered to the winds.

Another Frenchman, Gilles Garnier, also confessed to the crime in 1573. He told tales of devouring more than a score of young children, ripping them apart with his teeth and claws. Many witnesses were prepared to corroborate this. Garnier was burnt at the stake.

Perhaps it is not surprising, given the dire consequences, that many "turn-skins" wished to be healed of their affliction. One cure was to cut a werewolf's scalp. This happened by chance to a rich man in Palermo, Italy, who had wolfish tendencies. When the fits came upon him, his servant would let him out of the house through a secret door, so that he could roam freely. One moonlit night, he came upon a young man returning home from a drinking bout. The werewolf went for him, and the young man, too drunk to run, pulled a knife on him and slashed him across the forehead. As the blood poured forth, the wolf howled in agony and painfully resumed his human form. He never again came under the spell of the Moon.

Other unfortunates just tried locking themselves away on moonlit nights behind barred windows and locked doors. It was said that you could cure a werewolf merely by accusing him of being one, or by addressing him by his Christian name three times and drawing three drops of blood. Or, if the werewolf had entered into a pact with the devil, cutting the nail of his left thumb, which grows long and horny, would revoke the oath.

Is this all just the stuff of late-night horror movies such as "The Werewolf in London" (which was also a hit song)? Or is there more to it than we care to imagine . . .?

One suggestion is that the astral body materializes into a wolf, while the physical body remains in a trance. Another is that it is a form of mental illness in which a person runs on all fours, howling, and imagines himself to be a wolf. But what is not in doubt is the primeval fear it touches in us – fear of the wolf, of our own animal nature, and of the sheer uncertainty and terror of life itself. But whatever the truth about werewolves, why blame it on the poor Moon? Is it just because it is convenient to have something to blame, or does the full moon really affect us? The idea that the Moon can drive people crazy is an ancient one. And while full-blown werewolfism may be mercifully rare, more excitable, chaotic behavior at the time of the full moon has long been noted even among "normal" people. "I have a husband . . . but such a moonling as no wit of man . . . can redeeme from being an Asse."[63]

Does the light from the night sky also effect our psychic being in ways we are only just beginning to grasp?

Lunacy

The idea is even implied in the Lunacy Act of 1842, which refers to the various phases of the Moon as creating different mental states, with lucid periods occurring in the Moon's first two periods. An English law of two hundred years ago actually distinguished between the chronically insane and the "lunatic", who became deranged according to the Moon's phases. The keepers of lunatic asylums certainly believed in the Moon's powers: extra staff were put on duty at the full moon, and in the eighteenth century patients were flogged just before the Moon was full in some kind of attempt to stop them becoming deranged and violent.

> *"When the Moon's in the full,*
> *Then wit's in the wane."*

The very word "lunatic" derives from the Latin *luna*, Moon. The medieval physician Parcelsus called the brain the "microcosmic Moon", and he said that lunacy grew worse at full moon. This connection was taken for granted until about the end of the last century, special allowance was made for crimes committed at the time of the Full Moon.

Legends bear witness to the strange goings-on at the full moon – a Scandinavian fairy tale, *The Magic Mirror*, tells of King Alting, who behaved like a wild animal whenever the Moon was full. In Iceland, pregnant women are advised not to sit facing the Moon, lest their children be born deranged. Our English expressions such as "moonstruck", "moony", "mooncalf" and "loony", are evidence. The belief pervades our lives, however rational we think we are.

The Egyptians held that insanity could be cured by eating meatballs made from a particular snake, under the light of the full moon. It is interesting that the Egyptian god Thoth ruled both intelligence and the Moon. The Babylonian god, Sin, was also lord of both wisdom and the Moon. And in southern European countries insanity was thought to be brought by beings from the Moon. The association is, indeed, a long one.

There were many warnings to avoid moonlight, and especially not to sleep in it: the rational Hippocrates said that moonlight caused nightmares; the Talmud warns people not to sleep in it, and Plutarch declared that such behavior would result in insanity.

> *"The breach, though small at first, soon opening wide,*
> *In rushes folly with a Full Moon tide."*
> (Cowper, 1780)

Then, of course, there are Jekyll and Hyde. Apart from the Robert Louis Stevenson story that we all know, there is also the case of Charles Hyde, an English laborer, who was charged with criminal acts including murder. He was let off on the grounds that his behavior was triggered by the new and full Moons.

"What's this Midsummer-Moon?
Is all the World gone a madding?"[64]

It is claimed that there is more crime of all sorts at full moon. Is there any evidence for this?

Murders analyzed in Dade County between 1956 and 1970 (1,887 murders) showed a significant tendency for them to be committed at the time of the New or full moon. A similar trend was found in a 1978 study in Cuyahoga County. Other studies have not found this to be the case, but an interesting reason for this has been put forward by the usually skeptical H.J. Eysenck and D. K. B. Nias.[65] They point out that most homicide figures show the time of death, rather than the time of assault, and that the two may vary considerably. It just so happened that the Dade County figures showed time of assault. Crime statistics from Florida between 1956 and 1970 show that homicides peaked at the full moon and rose again during the new moon. A study in Ohio found similar patterns, but with a peak three days after the full and new moons. The difference was put down to Ohio's more northerly latitude.

The American Institute of Medical Climatology concludes that "crimes with strong psychotic motivation, such as arson, kleptomania, destructive driving, and homicidal alcoholism, all showed peaks when the Moon is full and that cloudy nights are no protection against this trend".[66] So it is not moonlight that is to blame – it is simply the presence of the Moon.

A new explanation is that the time of the local high tide (whether or not the place is actually on the sea), might be of more significance than the phase of the Moon. Studies have shown that when the Moon is highest in the sky locally – in other words, high tide – is the most significant time for crime.

When it comes to that old chestnut of full moon madness, how do we prove it? One way could be to plot the number of admissions to psychiatric hospitals according to Moon phases. But this is tricky – it may take several days for a person to be admitted. Even so, a study of over a thousand admissions to a mental hospital in Ohio did try to take this into account. And it found a significant coincidence between full moon and the breakdown.

What about measuring the effects of the Moon on the brain? It's been done: Dr Leonard Ravity, a neurologist, has taken measurements of the tiny electrical currents flowing along the nerves, using a micro-voltammeter. He found violent fluctuations at the times of the new moon and the full moon, and concluded that people who were already unstable became more so at these times.[67]

The Egyptians had a unique cure for werewolfism, eating snake meatballs beneath the light of the Moon.

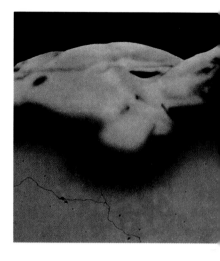

If all this is so, could not the Moon also affect our psychological powers? Our intuition tells us that, when that cool bright light shines full upon us, it gently brings forth those things which are hidden in the deep recesses of our beings. But in this age of science and skepticism, we must look for more concrete evidence. One intrepid scientist has had a go at measuring the elusive psychic and telepathic powers. In what sounds rather like the creation of Frankenstein, Andrija Puharich[68] describes his procedure:

"It took five years of preparation before I could complete a satisfactory series of telepathy tests for a full lunar month . . . The subject is Harry Stone, and six months of laboratory work had thoroughly disciplined him to control conditions...The Faraday cage was used for the constant environment . . . a matching card test was used throughout the experiments to evaluate telepathic ability. "Two peaks of scoring are evident, one on the New Moon phase, the second around the Full Moon phase. The latter shows the most pronounced increase in scoring rate."

Puharich puts this down to the same gravitational forces which affect the tides. The Moon "attracts" our telepathic powers, our kundalini energies, more strongly at certain phases. Perhaps we should not be surprised, given the links with menstruation, with werewolfism and the mysterious flow of life. Though an old idea, the field of research is still wide open. Let us end by mentioning some findings in the physical realm. That, after all, is something we can all believe in.

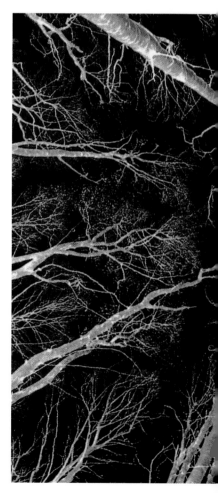

The Moon has been proved to influence bleeding. An American scientist, Dr Edson Andrews, found that eighty-two per cent of surgical bleeding crises occurred between the first and last quarters of the Moon, with a significant peak at the full moon. He concludes: "These data have been so conclusive and convincing to me that I threaten to become a witch doctor and operate on dark nights only, saving the moonlit nights for romance." (L. Watson, op.cit.) And if the Moon can influence bleeding, is there still not a story to be told about its effect on on our mental and spiritual life – the ebb and flow of our human tides?

> "Yes, lovely Moon! if thou so mildly bright
> Dost rouse, yet surely in thy own despite,
> To fiercer mood the frenzy-sticken brain,
> Let me a compensating faith maintain;
> That there's a sensitive, a tender, part
> Which thou canst touch in every human heart,
> For healing and composure."[69]

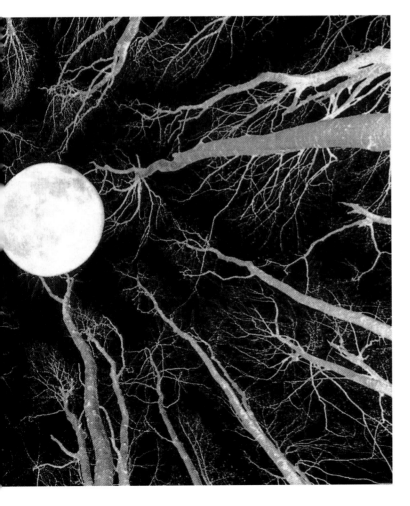

The Moon waxes and wanes, rises and sets, drawing the waters of the world with it. And, it seems, drawing us too, effecting us psychologically and physically.

Chapter 8

By the Light of the Moon

Rituals and Rites.

"Some worship for God the Sun, some the Moon"[70]

Turning a silver coin in the pocket at the first sight of the new moon; celebrating the harvest moon; curtseying to the Moon. These are some of the Moon rituals we have all at least heard of. But there are many more, to be found in all ages and all societies.

Some societies have gone as far as to worship the Moon: The Book of Job, probably the oldest book in the Bible, talks about Moon worship:

"If I beheld the sun when it shined, or the moon walking in brightness;

And my heart hath been secretly enticed, or my mouth hath kissed my hand;

This also were an iniquity to be punished by the judge: for I should have denied the God that is above."

This refers to the ancient practice of saluting the Moon, and the writer fears that the wrath of Yahweh will be invoked if this custom is continued.

The ancient residents of the Eastern Caucasus had a Temple of the Moon, in which they kept sacred slaves, many of whom were versed in the occult. When one of them showed more than usual signs of some kind of insanity, the high priest would bind him with a sacred chain and keep him in luxury for a year. At the end of this time he was sacrificed, by a sacred spear through the heart. According to how he fell, omens and portents would be read.

The ziggurat at Ur, a huge tower which dominates the area, was dedicated to the Moon god Nannar. The tower is one of the earliest ziggurats, built around 3000 BC. A sanctuary to Nannar was set up at the base of the tower, so that the god could descend through it and dwell in the temple. So heaven and earth were linked.

Others have turned to the Moon for help in times of trouble. In Eskimo communities, the Shaman puts himself into a state of trance, and "journeys" to the Moon to appease the goddess. During this time his body is cared for by his companions, while his soul encounters the dangers of the journey. His "return" is marked by public confessions of broken taboos which have offended the Moon who is, in east Greenland, lord of all animals.

Worshiping the Moon

There are many practices which come somewhere between religion and superstition. It is recorded that in the seventeenth century "the wild Irish" knelt before the new moon and said the Lord's Prayer. In Yorkshire they actually worshiped the New Moon on their bare knees. It was then still common for people to curtsy to the new moon, saying "Yonder's the Moon, God save her grace." Roman women "of the most noble and ancient houses" wore small crescent-shaped charms on their shoes, reported Plutarch. These were to catch the moonbeams which would otherwise enter their heads and cause untold damage.

A fifteenth-century author complained that "these days men do worship to the Sun, Moon and stars". In 1453 a butcher and a laborer from Hertfordshire were formally accused of claiming that there was no god but the Sun and Moon. One Richard Baxter, taking over a parish in Kidderminster in the seventeenth century, was shocked to discover that "some thought Christ was the Sun . . . and the Holy Ghost was the Moon."[71], while Ellen Green, an alleged witch, confessed in 1619 that "spirits came to suck her blood at certain phases of the Moon".

A Druid ceremony of sacrifice to the deities is described by Pliny. It had to take place on the sixth day of the Moon, and involved cutting mistletoe from

"These days men do worship to the Sun, Moon and stars."

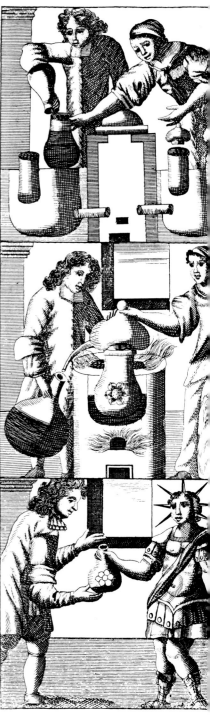

an oak tree: "Preparations were made for a feast and a sacrifice of two white bulls. A Druid in a white robe climbed the tree and cut with a golden sickle . . . the branch of mistletoe which was caught as it fell on a white cloak. The bulls were then sacrificed."[72]

Condensing the Moon

The ancient belief that the Moon was cold and watery led to a conviction that it was the cause of rain and dew. The alchemists tried to collect and condense moonlight, using large basins made of polished silver – a metal connected with the Moon, as gold is with the Sun. Hence the superstition that it is lucky to turn over silver coins in one's pocket when the new moon is first sighted. In Britain it was thought unlucky to point at the new moon when seeing it for the first time, and it was customary to raise one's hat or curtsy to it. If someone died at the time of the new moon, it was believed that more deaths could be expected soon in the family.

The moonstone is considered a sacred stone in India, and is traditionally only displayed on a yellow cloth, this being a sacred color. It is treasured as a gift for lovers, and is said to give them the ability to see the future. To do this, the stone is placed in the mouth at the time of the full moon. It has been claimed that some moonstones change color and markings according to the phase of the Moon. On special feast days an intoxicating drink called soma was made by the ancient Indians, and the plant which produced it was considered the king of plants and identified with the Moon. Soma was probably produced from hemp, and apparently had a similar effect to the modern-day Indian bhang drink, also made from hemp and well known to some followers of the hippy trail! The Moon was believed to govern all plants, and to favor especially the one producing soma.[73]

Heaven-Herding

Zulu heaven-herders are experts in reading the skies. To become a heaven-herder a novice must undergo a thorough preparation and initiation, which includes being cut, scarified and circumcized. The initiation – which may not sound too alluring, but is in fact highly prized – can only take place if the novice reaches the home of the heaven-herd at the first sign of the new moon. When the Moon is full, the initiate is considered also to be full of wisdom and knowledge, and ready to take his place as a heaven-herder.[74]

In Persia a statue of a lion was placed over the grave of a brave soldier. Cowardly soldiers were forced, on the night of the full moon, to pass several times beneath the body of the lion, in order to imbibe bravery.

The different phases of the Moon also have their own particular customs and rites. There are many for the new moon: it is welcomed back after the dark moon, and seen as a fresh beginning. "Grandfather, let us be at peace." Or, "Ah Moon, daughter of the air-spirit Deng, we invoke God that thou should appear with goodness. May the people see thee every day. Let us be." Such is the invocation of the Nuer of Southern Sudan at the sighting of the new moon. As they say this they rub ashes on their foreheads in the shape of a cross, and throw the ashes along with a grain of millet.[75]

At the time of the winter solstice, when the sun begins to regain the northern hemisphere, the Mossi of Upper Volta hold rites for the coming of the

The alchemists tried to condense moonlight, and the Moon was long believed to rule the growth of all plants.

Different phases of the Moon have been ascribed their own rites and rituals.

new year. They must be performed at the new moon closest to the solstice, the best time for opening the path of the Sun. The actual pathways leading to the royal palace are cleared of encroaching vegetation at this time; the king, having previously departed the palace, returns once the paths are cleared, just as the Sun will do. The winter solstice is also a time when some African tribes prepare their stores of medicines. But again, it must be done at the time of the new moon closest to the solstice. Swazi royal advisers scrutinize the morning and night sky to determine when the Sun and the Moon are in agreement.

Salute the Moon

The Dobu islanders of the Western Pacific look eagerly for signs of the new moon, as for them it symbolizes happiness. The children salute it by whooping – but this is not permitted while a man of importance is eating his evening meal! They also have a chant to invoke the new Moon's appearance:

> *"Over Kedagwaba the new Moon*
> *turns its back.*
> *They call crying on it*
> *They call crying on it*
> *They are happy*
> *They look out on the path*
> *They look for us*
> *They look for me*
> *They look.'*[76]

The Eskimos believed that if they brought snow into their igloos when the Moon was new, they would be helped in catching seals in the spring, because the rising Moon would give strength to the melting waters of the seas.[77] The Full Moon is a time of completion ceremonies. In Pagan times a lunar feast was held when the seed had been sown, at the full moon in March.

Japan's harvest moon is known as the "sweet potato moon". Sprays of susuki (eulalia) are laid out on the veranda to ensure a good harvest, as they resemble the rice plant. Dango, tiny skewered dumplings, are served, as well as the sweet potatoes which gave the time its name.

Moon-viewing has long been popular in Japan – there is even a special word for it, *tsukimi* – and Jugoya, the night of the full harvest moon of the eighth month, is the supreme one.

Great importance is attached to the full moon throughout the year in Japan. On the fifteenth day of the first lunar month there is the celebration of Koshogatsu, the Little New Year, and the important Bon Festival, the Buddhist Festival of the Dead, when ancestors are revered, falls on the fifteenth day of the seventh lunar month.[78]

Many religions honor the full moon as the "high tide" of psychic power. The Buddhists celebrate the Wesak moon, the full moon of May, because this was when the Prince Gautama was enlightened, while seated under a tree in meditation. The Hindu god Shiva is also often depicted meditating beneath the full moon.

The Greeks reckoned the day of the full moon to be the most propitious day for marrying. As Agamemnon said in Euripedes' *Iphigenia*, when asked on which day he would marry, "When the blessed season of the Full Moon is come".

Right: A period when the Moon is out of sight has its own quality. It is often believed to be a time when the unwary may succumb to illness.

Black and White Moon Medicine

The dark moon also has its own customs. The Babylonians thought that "the dark moon" was a time of great peril, and could only be overcome by fasting and religious ceremonies. The Tiv of western Africa warn that "the dark of the Moon" is when people are most likely to catch cold or be bewitched. The people of Bali used to use the time of the dark moon, in the ninth month, to expel devils from the island. The people would assemble at the principal temple, and offerings would be set out to the devil at a crossroads. Prayers were said to the priests, and a horn was blasted to summon the devils to the feast laid out for them. Men lit their torches from the holy temple lamp, then went through the streets with everyone shouting at the devils to depart. As much clatter and noise as possible would be made by everyone, followed by absolute silence which lasted for the next day.

It is not surprising that many rituals have sprung up around eclipses, even though they happen less frequently than the Moon's phases, for they are one of nature's most dramatic events. The Ojebway Indians thought that the Sun was being extinguished. They shot fire-tipped arrows into the sky, hoping to relight it. The Sencis of Peru did the same thing, but to chase away a wild animal which they thought was fighting the Sun. The Arawak of Guiana believe that the Sun and Moon are fighting when the Sun is eclipsed, and in order to separate them, they utter terrifying cries.

During eclipses of the Moon, some Orinoco tribes would bury lighted brands in the ground, saying that all light on Earth would go out if the Moon was extinguished, apart from any that was hidden from sight.

The Chilcotin Indians would tuck up their robes as though for traveling, and, leaning on sticks as though heavily burdened they would walk in a circle until the eclipse was over. This was to help the Sun as he walked wearily through the heavens.

Similarly, in ancient Egypt, the king, the representative of the Sun on Earth, would walk round the walls of a temple to ensure that the Sun maintained its steady progress through the skies without eclipse or other mishap.

As we have discussed, the Moon has quite a reputation in the fertility stakes, so here again there is rich ground for moony rites. One South American ritual is supposed to endow a man with a large penis. When the Moon is full, young Mocovi boys pull their noses and ask the Moon to lengthen their masculinity. Another ritual requires the participation of other people; the ancient Araucan believed that the Moon in its various phases personified a young girl, a pregnant woman and an emaciated old woman. At the time of the full moon, dancing men would tie a finger-thick rope of wool to their penises on which the women and young girls pulled. This rite was, according to accounts, followed by "promiscuous scenes".[79]

The traditional Tswana settlements in Africa are structured using the Moon as a model. At first the *ktolga* (village) may seem to be growing haphazardly, but as it progresses it forms a crescent, like the Moon as it begins to wax. The Tswana correlate this with the growth of the family; both must grow like the Moon until they become full. When the circle of the settlement is closed, again like the Moon, it is time for another to begin.

A Bantu father may not hold his offspring until the child has passed through the *yandla* ceremony, an important rite of passage. At the first new moon after the resumption of the mother's menstruation, the mother takes a burning torch and hurls it at the Moon. The grandmother tosses the baby in the air, saying, "There is your Moon." She lays the child on a pile of ashes, and only then is the father permitted to take the baby in his arms. The ceremony symbolizes the baby moving on from the mother towards the father. Moon and ashes symbolize a cooling; the child is moving from the warm maternal world to the cool, social world of the father, at the time when the Moon itself is also in its infancy.[80]

The African Lakher claim that ill-feeling between a woman and her brother or maternal uncle will result in her infertility. To cure this, when the Moon is waning either of these men places some fermented rice in the woman's mouth with a hair pin, and the two do not speak again until the new moon has risen.

It is said that a man's virility can be enhanced if certain rituals are practiced at the time of the Full Moon.

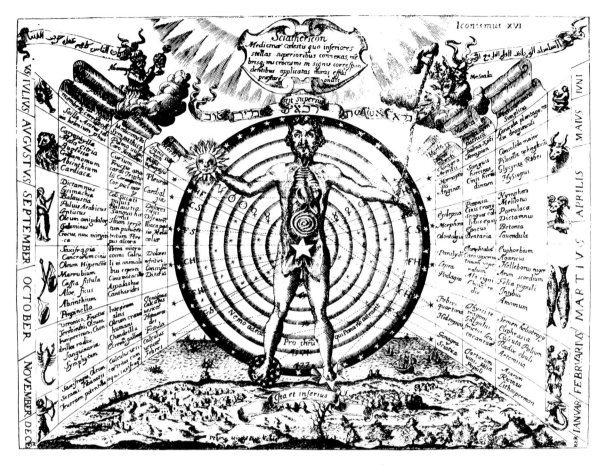

In medieval medicine, each part of the body was believed to be governed by a particular sign of the zodiac; the crayfish represented the Moon's influence.

Of all the customs and practices to do with the Moon, the ones most familiar to us today are probably the medical ones – what we would call folk remedies. Some of the cures, though, sound worse than the illness!

The Thonga of Natal, South Africa, have an elaborate ritual for ridding a person of possession and illness. They are cured with blood, sweat and probably a few tears – but not until the Moon is right! They believe that illness happens when a person's relationship with the Moon has gone awry, so the rituals can only begin at new moon. The ceremonies last many days, and include a wild drum performance, in which the "patient" may throw himself into a fire with no apparent harm resulting, followed by the immersion of his head, with eyes open so that he may see anew, in a bowl of water. Some time later the blood of a chicken or, preferably, a goat will be drunk. By the time the body has been cleansed, man and Moon are restored to harmony, and the man is again himself.[81]

There are many customs and practices in folk medicine which make use of the Moon as part of the "prescription". One Swiss cure for illness is to cut the fingernails and toenails on a Friday night when the Moon is waning, then force the parings under the shell of a crab, which will take the fever away with him. Shingles could be cured, according to a relatively recent custom in America, by the blood of a pure black cat, especially when killed by the light of the Moon. The blood should be applied to the skin for several hours. Documentary evidence verifies that this actually was practiced. Ringworm was treated in old Los Angeles by spitting on a gold thimble, placing it on the infested area and turning it three times – all by the light of the Moon. In the New Forest, Hampshire, sick people would hang holed stones around their necks to cure themselves. These had first to be exposed to the light of the full moon for three nights.

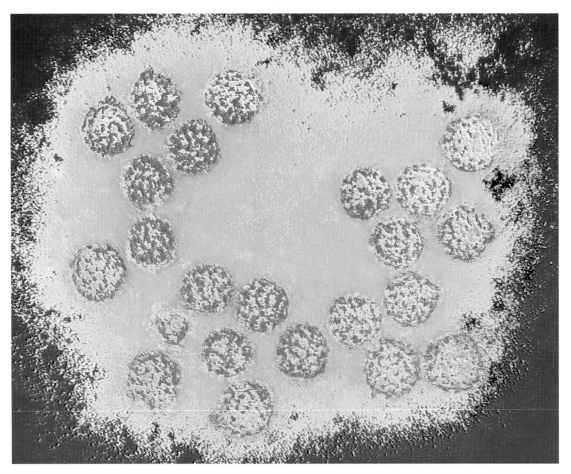

There are many cures for warts which use the Moon. For instance, if you happen to have a corpse handy just as the Moon is beginning to wane, try rubbing your hand three times over the dead body, then touch the wart. A less dramatic method is used in New Hampshire; at full moon the wart is rubbed with a bean, which is then thrown over the shoulder. In Texas, of course, they do everything bigger – they throw a dead cat over the limb of a tree by the light of the full moon. Illinois keeps things simple – when the New Moon is first seen, you bend down and pick up whatever happens to be underfoot, then rub it on the wart and throw it over the left shoulder.

A rupture in a child can be healed with the help of the full moon, according to San Diego folklore. A young willow tree is cut in half lengthwise when the Moon is full, and the child is passed through it. The two parts of the tree are then tied back together, and as they rejoin, so the rupture heals.

The pestilence could be staved off if twelve naked youths and maidens ploughed seven furrows around the village, from midnight on the Saturday and Sunday after a new moon.This was the practice in southern Slavic countries, where it was forbidden to speak to, touch or look lasciviously at the plough people. It is said that in old Hungary on the night of the New Moon, the crippled would roll in the dew, and the blind would wash their eyes in it.

Some medical Moon advice is even more specific. Dental cavities, for instance, should be filled in the third or fourth quarter of the Moon, when it is waning, and in a fixed sign of the zodiac (Aquarius, Taurus, Leo, Scorpio). Teeth should be extracted during a waxing Moon, but only when the Moon is in Pisces, Gemini, Virgo, Sagittarius or Capricorn. Even dental plates have their own time apparently – they should be made under a waning Moon, in one of the fixed signs. To remove unwanted growths, such as corns, hair or warts, use a barren sign (Aquarius, Aries, Leo, Virgo or Sagittarius), with a fourth-quarter Moon. But for surgical operations wait until the Moon is waxing, in the first or second quarter, when wounds will heal faster.

Let us end with a custom from Naples. Women call on the fertile power of the Moon to increase the size of their breasts. They stand totally naked in the moonlight, and recite an incantation nine times: "*Santa Luna, Santa Stella, fammi crescere questa mammella*", meaning "Holy Moon, holy star, make this breast grow for me". The growth of the breast is thought to coincide with the growth of the Moon itself. Tests have been made on this last one, and it was found to work in ninety per cent of cases. Whether that is just the result of giving them extra attention or because of the Moon's influence is for you to decide!

Above: Of all the rituals connected with the Moon, the most common are to do with its effect on our health.

Left: Cures for warts which depend on the influence of the Moon are to be found throughout the world.

Part II

OVER THE MOON

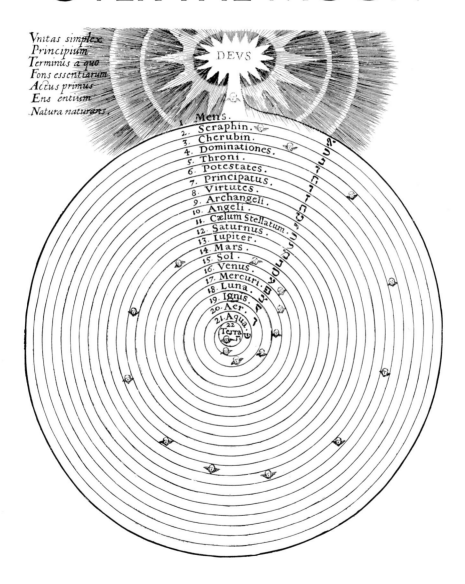

Chapter 1

Moon Stones

We have seen how the Moon has been held in awe and reverence throughout the ages. Man has created Moon goddesses, and performed rituals and ceremonies in her honor. In this section we look at the tangible evidence of the Moon's influence in our lives, at how we have incorporated it into our day-to-day existence.

All around the British Isles, Europe, Central and South America there are ancient stone circles and megaliths. For centuries they were dismissed by most people as meaningless piles, but while they were being broken up to make millstones or walls, their mysteries were still kept alive by some, even if they only sensed their significance as a distant echo.

Today, many people are trying to amplify and listen to that echo. And through the mists of time the voice of the Moon is heard yet again. Perhaps a sixteenth-century Scottish historian knew a thing or two when he called the stones: "the old temples of the gods" in which "the New Moon was hailed with certain words of praise". We will never know what these words were, but science is now showing us more and more about the stones' connections with the Moon.

So what have these piles of stones got to do with the Moon? Some of them, as we will see, are astronomical calculators of some kind, and have been constructed so that they can predict important movements of the Moon.

But there is another, more mysterious, link with the Moon. Many groups of stones just bear witness to the importance of the Moon in the lives of the people who constructed them. The Moon, it seems, was at the heart of their spiritual beliefs. Many stones are aligned with the Moon's uppermost and lowermost points in the sky: could these have been for ceremonial purposes? Perhaps they wanted to "catch" the Moon's essence as it traveled through the rings they had built to mirror the Moon's own circular movement.

What makes us think that the Moon was of such great importance to the people who made these structures? We should try to put ourselves in a place where no street lights tempered the pitch dark of a moonless night, where a

Astronomers believe that many ancient standing stone formations were used as observatories.

lamp could not be switched on at the flick of a finger, where moonlit nights assumed a far greater importance than we will probably ever know. And the Moon's monthly cycle was a far easier to see and more effective measure of time than was the Sun's slow journey over a year.

Let us take a look at some of the sites which seem to have been constructed very much with the Moon in mind.

On a bleak shoreline, a circle of tall thin stones appears on the horizon. These are the Standing Stones of Callanish in the Hebrides, islands to the north-west of Scotland. The bleakness of their setting adds to their mystery; just to survive in such a place must have been hard enough, why use so much energy to build this?

Astronomers now believe that the ring, with its dominant central pillar, may have been used to predict eclipses, and to make calculations about the Moon. Avenues of stones extend towards the compass points from the ring, and the stones may have been aligned towards important movements of the Sun and Moon.

But another mystery remains. A century ago, cremated human bones were found beneath the central pillar. Was this a ritual burial connected with the Moon? We can only speculate on the circumstances of the death, but it looks as though ancient man wished to make a powerful statement about his connection with the Moon, the skies, the mysteries of life and death.[83]

There are rows and rows of standing stones at Carnac, in Brittany, an amazing and still mysterious configuration that extends for literally miles. What purpose could they possibly have had? Even now, nobody really knows. What is almost certain is that one of the nearby monoliths, Le Grand Menhir Brise at Locmariaquer, was almost certainly used for lunar calculations. This stone is now split into four, but must originally have weighed at least 355 metric tonnes, and stood twenty meters high. It may have served as a foresight marker for the Moon's rising and setting at the four major extreme points of its 18.6-year cycle. Eight observing points would have been needed – four for the rising, and four for the setting of the Moon. On investigation of the monolith it was found that at no fewer than four of these theoretical points, prehistoric markers – mounds or stones – still exist.

The Merry Maidens is a wonderful circle of small stones at Land's End, Cornwall on the south-west coast of England. A group of girls were said to have been changed into stone for dancing on the Sabbath. As in most of the Land's End circles, the tallest stone is at the west-south-west, meaning that it could have acted as an astronomical marker on the minimum moon.

Many neolithic stone circles originally had a cove – three large standing stones – at their center, which faced the maximum midsummer setting of the Moon. Such a one is Arbor Law, Derbyshire, England, although the three stones are now collapsed. Again there are associations between the Moon and death – the coves were like the entrances to chambered tombs where funerary

Above: The Merry Maidens at Land's End, England.

Below Arbor Low, Derbyshire, England.

rites were performed. At Arbor, a man's skeleton was found buried alongside the cove. These magnificent circles and alignments of standing stones from about 2500 BC, found in various parts of western Europe, such as Avebury, Stonehenge, Arbor Low and Stenness, still hold many mysteries. They are all far bigger than they would have needed to be for purely observational purposes, suggesting that they also had some kind of ceremonial function.

Silbury Hill, near Avebury, England was probably aligned to both Sun and Moon, with its most important role being at harvest time. We can only speculate on what may have happened at that time, but the ancient goddess/Moon symbol of a pair of six horns has been found embedded in the earth there. In the sanctuary at Avebury, a circular shaft pointed to the burial site of a fourteen-year old boy, all his bones broken, lying in a fetal position on his right side, facing east and with the bones of a young ox on top of him alongside a horn amulet. Archaeologists believe such sacrifices might have been made at the beginning of winter to appease the dying Mother. Since the Moon was of more importance than the Sun in the agricultural calendar, such ceremonies would probably have taken place on a night of the black Moon.[84]

The Doubler Stones in West Yorkshire, England, are covered in prehistoric carvings and cup-marks which may symbolize the Moon.

Kit's Coty House in Kent, England is a dolmen, three upright stones topped by a capstone. A time-honored ritual performed there initially appears pointless, but may be the remnant of a more ancient rite. If, at the time of the full moon, an object is placed on the capstone, and the petitioner walks thrice around the stones, the object is said to disappear. But why should anyone want to do this? Perhaps the fact that the dolmen was originally part of a neolithic burial chamber suggests that there was once a sacrificial ceremony there, of which this is now just a distant echo.

Before instruments for observing the Sun, Moon and stars were invented, the only way to measure their positions was by noting where they rose and set on the horizon. A simple way of doing this would be to mark the Moon's setting point by a conspicuous natural feature on the horizon – such as a prominent mountain peak. Two standing stones would then be set up, permanently marking the observervation point and showing the position of the horizon mark. The line from the stone (the backsight) to the horizon mark (the foresight) is the alignment. This way, even the complicated movements of the Moon could be used to predict eclipses.

Moon observatories probably began to be built around 2800 BC, the time at which the first structures at Stonehenge were made. Most the megalithic monuments in Britain and Brittany could have been used as lunar observatories, as well as for ritual purposes. These observatories came in all sorts of shapes and sizes: as well as the stones and circles, there were large earthwork avenues called a cursus (there is one near Stonehenge, and a six-mile long one in Dorset).

These align at various points with horizon markers indicating the rising or setting points of important phases of the Moon. Single menhirs could be used in the same way, when aligned with barrows or with marks dug in distant hills. This interest in marking the Moon's progress was no whim. It was an inbuilt function of these structures for about one thousand years.

What of that most famous of stone circles, Stonehenge? Could even that have had its lunar aspect? In the 1960s, Professor Gerald Hawkins' computer calculations revealed that as well as being aligned for the summer solstice, Stonehenge had other significant alignments, to the rising and setting of the Sun and Moon at equinoxes and solstices. And it was clear that Stonehenge could have been used as a computer for predicting eclipses.

Stonehenge is famous – even notorious – these days for the celebration of the midsummer sunrise. The Heel-stone, when seen from the center of Stonehenge, marks the place on the horizon where the midsummer sun rose in 2900 BC. What is less well known is that it was probably also the marker for observing the Moon.

This has been worked out by century-long sightings towards the shifting positions of the moon-rise, which, before the ditch and bank were built, were recorded by rows of stakes. They were called the Aubrey Holes after John Aubrey who discovered them in the seventeenth-century, and were markers for extreme northerly moonrise positions over a period of at least a century.

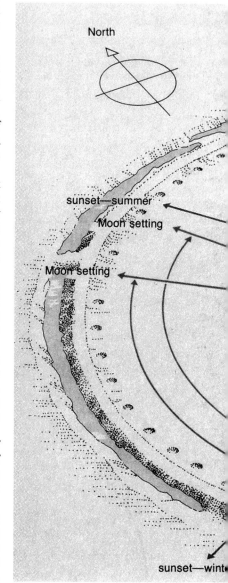

Stonehenge acts as a giant observation station of the heavenly bodies. The alignment with the summer solstice is well known, but it is now clear that alignments with the Moon are equally important, and possible date from an earlier stage of construction.

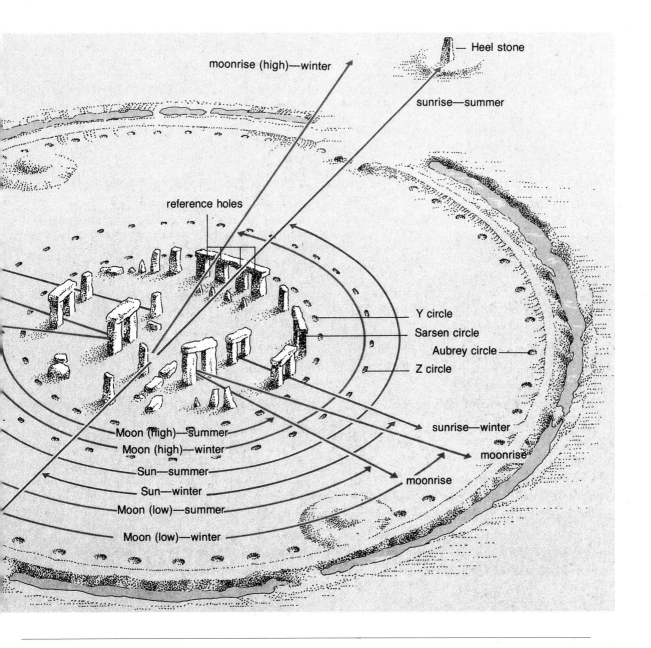

moonrise (high)—winter

Heel stone

sunrise—summer

reference holes

Y circle

Sarsen circle

Aubrey circle

Z circle

sunrise—winter

moonrise

Moon (high)—summer

Moon (high)—winter

Sun—summer

Sun—winter

Moon (low)—summer

Moon (low)—winter

moonrise

 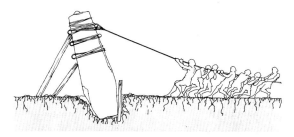

Deep pits with one sloping side were dug to house the great sarsen stones, some of which weighed as much as 45 tons. The stones were probably lowered into the pits with the aid of ropes and rollers, then slowly raised by using a scaffold as the fulcrum of a rope-operated lever. In the last stage, ropes alone may have been used to straighten the stones.

They show where the winter Moon rose during a fifty-six-year period which takes into account all the different positions in its 18.6-year cycle. The Moon rose between the two right-hand posts, a quarter of the way from the midpoint of the cycle to the major standstill, and between the two left-hand posts when the Moon was halfway to the major standstill. The Heel Stone and a row of smaller stones acted as outlying markers. The Moon, at its maximum, would have risen well to the north of the midsummer sunrise position as marked by the axis of Stonehenge, but nine years after the maximum, half-way through the lunar cycle, the two positions would have coincided. So it seems that Stonehenge was designed as a kind of stage for celebrating the two polarities of the midwinter moonrise at night and the midsummer sunrise at daybreak – a beautiful duality.

Perhaps – and we can only speculate on this – the fact that these Sun and Moon lines coincide in this way suggests that there may have been a special celebration every eighteen or nineteen years when the two were in accord. For, although the system was accurate enough to tell where the Moon was in its cycle, it was not very precise, and it seems that the posts were erected so that the whole community could see the moonrise as part of a ceremony.

Could there have been a symbolic connection between the Sun, Moon and death at Stonehenge? The circle may have originated as a timber hut in which corpses were left to rot, then the later Grooved Ware people built a henge around it to bury their cremated dead. Next, the Beaker people put up a stone circle; this was pulled down by later people, who set up the gigantic sarsens which are still there today. In other words, Stonehenge was originally a temple. Certainly, most of the burial mounds around Stonehenge have their burial ends facing either the extreme moonrises or sunrises, or they face east where both the Sun and the Moon rise in spring and autumn.

Before the stones arrived, could Stonehenge have originally been a Moon observatory, dedicated to some form of the Mother Goddess? It is not impossible: a pair of ox horns – an ancient symbol of fertility and the Moon – have been dug up in the central area. Some kind of priest, or magician, could at some time have occupied the central position, surrounded by a maze of timber markers, "painstakingly fixing the realms of sky and earth together in an everlasting moongate. In the slow course of time the wandering Sun and

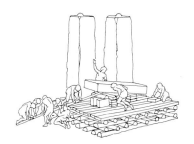

Once the uprights were in position, the 7 ton lintels were probably placed on a timber platform. As the stone was inched higher with levers, wedges and blocks, the platform

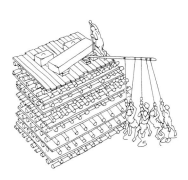

was built up from below. Finally, the lintel was slid into position on top of the uprights, which had pegs carved to fit exactly into the mortise sockets on the lintels.

Moon and the invisible forces of the underworld were to be harnessed inescapably to the will of man."[85]

Archaeology is finding more and more evidence of how these sites are linked with Moon observation. At Newgrange, an Irish megalithic mound which contains a passage and chamber, an archaeologist recently found that there were astonishing alignments with the winter solstices – the Sun actually strikes some carvings on the very day of the solstice. Armed with this discovery, he set out to find whether the full moon also played a part:

"We sat in total darkness. We were surprised by the sudden appearance of a mysterious thin shaft of silvery white light which penetrated the passage and created a glowing patch of brightness on one of the chamber stones. We watched in amazement as the size of the patch of light diminished rapidly until it disappeared before our eyes. It was like an apparition . . . singularly the strangest and yet the most beautiful event we had witnessed inside a passage mound. After that I never doubted that, whatever the achievements of the megalith builders in the field of astronomy, the structures they built had a strong element of ritual observation in their function."[86]

Brennan calculated that it was virtually impossible for such falls of the moonlight to be purely coincidental, and he found a 18.6-year calendar appearing. Perhaps the engravings in the mounds, of spirals, concentric circles and lozenges, also celebrate the Moon. And could the fact that the long passage was aligned to the winter solstice suggest that it may have represented the birth canal, new life after the solstice?

At the very least, the incredibly accurate spherical geometry of the passage graves, Avebury and Stonehenge, and the patterns on passage grave walls, would suggest that they must be inspired by the arrangements of the stars and the planets, of which our forebears would have been so aware. It is curious that all the stone circles, standing stones and avenues can actually be distinguished most clearly from their surroundings when seen from above – just as the incredible patterns made by the Nazca Indians of Peru do in the desert, appearing as images when seen from the sky. Maybe man saw the Earth as part of the vastness of the universe, and was offering up the monuments as a homage for the Sun, Moon and stars to behold. In fact many megaliths are inscribed with "cupmarks", circular depressions which have been ground into the rock. They are often associated with burials, and may be symbols of the Moon. One example of many is at Kittierney, County Fermanagh, where a recumbent stone with cupmarks is aligned on the midwinter moonrise.

Another hint of the Moon goddess has lingered in the language of the stones: a lot of standing stone formations include the number nine in their names – such as the Nine Maidens – even though they may not actually have that number of stones. It could be that the Great Mother Goddess, as embodied in the Moon, was worshipped here. The Goddess had the three attributes of new moon, full moon and waning moon, which eventually became three goddesses. These in turn were thought each to have three aspects – thus making nine goddesses, or maidens.[87]

Ancient figures carved into the hillsides of Britain may also have Moon connections. The Gogmagog Hills near Cambridge are carved with the figures of Gog and Ma-Gog, said to represent respectively the Sun and the Earth or Moon goddesses. Ley-line fans also claim that the figures are on a line linking various prehistoric barrows and forts.

And what of the stones themselves? If you have ever visited these places, you may have felt drawn to reach out and touch the stones, as though they contain some charged energy. Local legends have evolved around many of the ancient stones, telling of them moving as though they really did have some kind of life of their own. But usually, according to the tales, they needed some kind of catalyst to do so, and this was often the Moon. The Waterstone at Wrington, Avon was said to dance, but only when the full moon fell on Midsummer Day.

Rituals involving the stones and the Moon also grew up over time. The Stone of Odin in the Orkneys (sadly destroyed in 1814) was highly venerated for its ability to impart magical powers. Those wishing to receive such blessings had to visit it at the full moon in nine consecutive months, crawling around the stone nine times on bare knees, then making a wish while looking through the hole in the stone.

The stones on these ancient sites are said by dowsers to be affected by the Moon, possibly through the medium of underground waters.

Dowsers say that the stones seem to "channel and be surrounded by a helical sheath of force"[88], which changes according to the Moon's phases. Could this be connected with the fact that the Celtic calendar was lunar, each month beginning on the sixth day after the new moon, and each year on the sixth day of the first new moon after the vernal equinox? One dowser even claims that the stones alternate between being positively and negatively charged, and that the polarity changes with the lunar cycle exactly in accordance with the Celtic calendar.

Right: Human remains have been found at several sites of ancient stone formations. The positioning of the stones suggests they may have been human sacrifices to the Moon.

Chapter 2

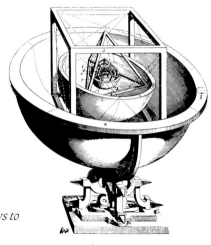

Moon Signs

"Everyone is a moon and has a dark side which he never shows to anyone."[89]

"   I do not know all the influences which go from body to body, I do know
    that if man is not affected in some way by the planets, Sun and Moon,
    he is the only thing on Earth that isn't."[90]
We may no longer be building stone circles to bring the Moon goddess down
to Earth, but we still invoke her influence, and seek to find our mysteries
reflected in her. The gentle arts of astrology and tarot know a thing or two
about the connection between us and the Moon.

Astrology

In astrology today, the Sun is given predominance over the Moon – we all
know our Sun signs, the twelve signs that appear in the daily paper, but how
many of us know our Moon sign? The Moon's position in your chart is of great
significance: after all, it goes into a different sign roughly every two and a half
days, while the Sun only changes its sign twelve times a year. But the Moon
has not always been so neglected.

"The divine and light-giving Moon, waxing in crescent, was running in
Taurus 13 degrees and a thousandth part of a degree; in the sign of Venus; in
its own exaltation; in the terms of Mercury; in a female and solid sign; like
gold; mounting the Back of Taurus." (From one of the earliest surviving
horoscopes, Greece, AD 81)

The ancient Babylonians had a sophisticated knowledge of astronomy, and
worshiped the Moon as the queen of the night. The Earth was thought to be

her child, and hence still under her influence. In ancient Rome, the Moon sign was thought to be of more importance than that of the Sun, and people's signs were classified according to this.

Before Copernicus set matters straight, the Earth was generally believed to be the center of the universe (see illustration). Around the Earth are various spheres beginning with Saturn and ending with the Moon. The Moon was supposed to draw the hidden powers of the stars down to Earth. Each of the planetary spheres was said to be governed by a particular "intelligency" – the one for the Moon was the Angel Gabriel.[91] He is sometimes shown with a crayfish, the ancient symbol for Cancer which is ruled by the Moon.

Many Medieval illustrations show the Moon connected with woman's womb or genitals. Medieval astrology used the Moon, rather than the Sun, to study the pathway through the fixed stars, because it moves through the

The Angel Gabriel is seated beneath the crayfish, symbol of the Moon. He was held to be the "intelligence" presiding over that planet.

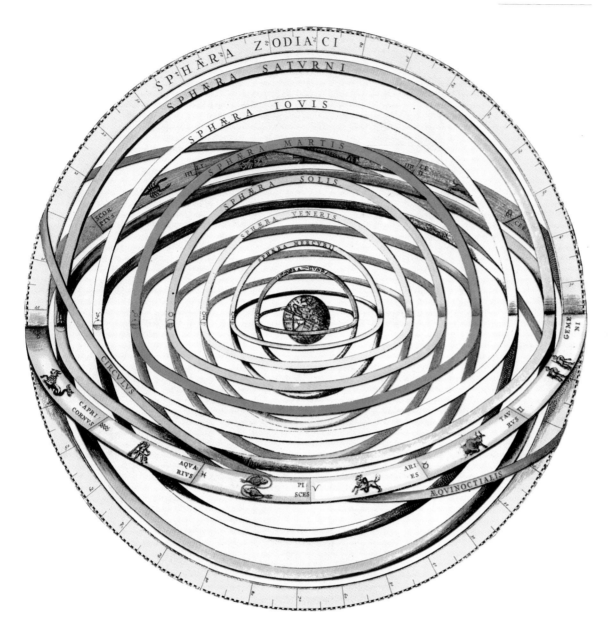

In the pre-Copernican universe the Earth was
thought to be the center of everything.

zodiac roughly every month rather than every year. They divided the skies into twenty-eight, approximately the days of a lunar month, and this was known as the lunar zodiac. The main stars within each of the days were called the stations of the Moon.

In the sixteenth and seventeenth centuries, popular almanacs told people how to conduct their lives according to the phases of the Moon. There were many activities which it was considered best to do according to Moon phases. These almanacs would tell you, according to the Moon, the precise day on which to move house, get married, travel, purge yourself, let blood, even cut your nails! But their recommendations were probably about as accurate as the Sun-sign horoscopes found in our daily newspapers, and certainly equally as popular! What did these ancient astrologers see in the Moon? They considered it to be a symbol of man's understanding, its phases representing the gradual increase in his perceptions. The Moon reflects the countenance of the Sun, as a quiet pool reflects the trees and the sky with a gentle clarity. Thus man may look upon the Moon in order to perceive the Sun's activities, for to gaze directly upon the Sun is dangerous.

If ideas about the Moon's effect on us have faded in more modern, rational times, they have never disappeared completely. Madame Blavatsky, founder of the Theosophists, kept them alive in the nineteenth century.

"If certain aspects of the Moon effect tangible results so familiar in the experience of men throughout all times, what violence are we doing to logic in assuming the possibility that a certain combination of sidereal influences may also be more or less potential?"[92]

She quotes the Hindus of Travancore, who say: "Soft words are better than harsh; the sea is attracted by the cool Moon and not by the hot Sun."

"Born with the moon in Cancer
Choose her a name she'll answer to
Call her green and the winters cannot fade her
Call her green for the children who have made her
Little green, be a gypsy dancer.[93]

As to what the Moon means in your own personal horoscope, the subject is a complex one, towards which whole volumes have been devoted. The Moon changes sign about every two and a half days. You can find out which sign your Moon is in by consulting astrological ephemera. But even then, to get a true understanding of its meaning for you, it is advisable to have your chart drawn up by an astrologer, who will be able to see how the Moon relates to the rest of your chart or alternatively there are many excellent books on the subject. We can, though, look at what the Moon generally represents in astrology. The elusive Moon goddess, not surprisingly, has come to represent the feminine or "yin" part of the psyche. It governs in the sign of Cancer, the Crab, and rules the life of the unmanifest world – the unconscious, the spirit, but also the maternal qualities of conception and birth. Its mercurial, silvery light calls forth the nocturnal, subterranean parts of our selves – the occult, magic – and of course the Moon irresistibly attracts the light of the Sun, the male principle, to her. The Moon is the *anima*, linking man's conscious ego with his inner spirit. The Moon in the chart is also said to symbolize childhood, a person's early world, the kind of home in which they grew up and their relationship with their mother. It shows too their most basic, instinctive kind of behavior, and how this manifests itself in close, emotional relationships, and in any other situation where instinct comes to the fore.

So the Moon represents two archetypes in astrology – there is the outward appearance, the way of life we are most comfortable with because of the way we have "reflected" our environment, particularly in childhood. And there is the more mysterious link with the source, the origins of life – if you like, the Great Mother. Maybe this is why the astrological symbol for the Moon is the crescent: the two joined bands bring together matter and spirit, conscious and unconscious, and show each of us how to resolve the duality.

The Moon is said to rule the body's fluids, in particular the semen, with its white, pearly quality like the Moon itself. It also rules over growth and nature's hidden processes, those things that take place in the hidden parts of the Earth.

According to ancient astrology, the Moon in your chart is in the same place as the Sun was in your previous life! So if your Moon is in Virgo now, your Sun-sign was Virgo in your last life. An interesting speculation. The Moon sign can be seen as a link with your past life in terms of childhood and heredity. It shows where your deepest needs lie – what has nurtured you, what still gives you security, and what umbilical cords you need to cut.

Above: In our hectic, technological era it is hard to spare time to look towards the heavens, and yet we feel an increasing yearning to feel the gentle pull of the Moon.

Michel Gauquelin, renowned for his extensive scientific research into the claims of astrology, has found that if one or more parent was born when a planet was just past the rising point or midheaven, there was a significant chance that their children would also be born under this particular planet. This applied particularly to the Moon, Venus and Mars. Incidentally, he also found that this did not apply to induced births, only to natural ones.

The different phases of the Moon, and how each affects our behavior in day to day life:

New Moon: beginnings, hidden changes, chaos and disorganization, confusion, rest

Full Moon: completion, fulfilment, activity, unrest, awareness

These can be further broken down:

1st quarter: beginnings, outward-going, germination and coming forth.

2nd quarter: the development of things which have already started

3rd quarter: completion and maturity, fullness

4th quarter: rest and introspection, disintegration before new beginnings

To work out the placement of the moon in an astrological chart one would need to use an Ephemeris (a book which maps out the movements of the planets from day to day) or a computer with astrological software programmed in.

Using an Ephemeris, the first step is to find the date of birth or day which you are interested in, then look up the moon sign. The Ephemeris will give you a reference such as 14 degrees (please use sign here) 15′ (G.M.T.) at noon or midnight depending on which book you are using.

To calculate the position of the moon for an American location you must either add or substract hours according to G.M.T. Because the Moon is the most rapidly moving planet (moving 1–6 degrees every 2 hours) it is important to be as accurate as possible with your calculations.

The moon in an individual's chart represents the feminine aspects of one's nature; the way one feels and senses, the childhood, the mother and the unconscious. In the following pages we will explain what it means to have the moon in each of the 12 astrological signs.

People with the Moon in ARIES are very active and aggressive with a love of the self. They are dynamic, make good leaders and possess an adventurous spirit. Generally the challenge for these people in life is to have the courage to go for what they want. The negative personality traits are a slight excess of self interest and an unwillingness to work or co-operate with others. Understanding inter-dependence and how to relate to others on an equal level will rescue them from this negative trait in their character.

Individuals with the Moon in TAURUS tend to be quite stubborn, but very sensual, with a love of the finer things of life; food, material wealth and beauty through which they physically nourish themselves. They are loyal, supportive and industrious. Trust in their personal survival and learning to value their feelings are formative challenges as the Moon-in-Taurus people tend to worry about these issues.

GEMINI Moon people are interested in everything. They are fickle, intuitive, imaginative and make very good mimics. These people tend to approach life intellectually; the challenge for them could be to become a little more spontaneous and playful with their feelings and to communicate more from the heart, introducing thus a balance between their thoughts and feelings.

The person with Moon in CANCER tends to be emotionally insecure and moody, and does not like to leave the home base. The challenge is to feel emotionally more secure and to gain trust in love. Their positive attributes are their loving character, their care for people, and they might have a special skill in the area of healing.

The person with the Moon in LEO is gifted, creative and emotionally dramatic, with a love for playing games and sports. They are very much identified with their ego role, and possess a desire for recognition from others; they also love being in control. They would do well to get more in touch with loving their own beauty and not always looking for that love from others.

Those born with the Moon in VIRGO have an earthy innocence, and love being alone. They are precise organisers, but tend to act out of duty and martyrdom. Their challenge would be to overcome their negative attitude towards feelings and accept that they are perfect as they are.

With the Moon in LIBRA you will find artistic, musical people who are loving, charming and like to relate to others. They tend to be very diplomatic and often can become dependant on others. Their challenge is to develop a peaceful relationship with their own feelings.

The SCORPIO Moon endows those born under its influence with highly psychic and intuitive powers and a deep understanding of other people's feelings. This attribute in an unconscious person might be used to manipulate others. These people are the classic "prima donnas" and are often caught up in intense emotional dramas, which could be avoided by a simpler attitude towards their emotions.

Those born with the Moon in SAGITTARIUS are optimistic, enthusiastic leaders who love adventure and travel. They are inspiritional teachers who find it easy to encourage and uplift people. As they tend to experience life with the purpose of gaining knowledge, their challenge is to learn to experience life more directly through their feelings.

The person with the Moon in CAPRICORN has usually had a traditional or restrictive childhood in which parents played an important role. This caused them to be serious children, with a tendency to feel separate and aloof from others. In adulthood, they develop strong feelings of responsibility and duty and make great organisers who like being in control. These people tend to become lighter and more playful as they age. Their particular challenge is to be able to respond in a relaxed and clear way, without any of the tensions that the feeling of carrying the world on their shoulders provokes.

With the Moon in AQUARIUS you will usually find quite eccentric, revolutionary individuals. They may be gifted with computers or astrology and these may well play an important role in their lives. They hate being told what to do. Through idealism they can become dispassionate and cold, so their challenge is to feel and express their emotions in their own unique way without condemning others. People with their Moon in this position can be very good spiritual channels.

People with the Moon in the last placement on the astrological wheel, in the sign of PISCES, tend to be very loving, giving and compassionate, with a tendency to go overboard and become victims. They can also be a little tricky and deceptive, because of their sensitivity and sometimes it is hard to pin them down as they float along in a rather nebulous world, being prone to 'bliss' attacks. Their challenge is to trust their intuition and to trust life.

THE HIGH PRIESTESS

Right: The Moon in tarot hints at the mysteries of the unconscious.

Tarot

Tarot is another ancient way of tapping into the mysterious flow of our lives. The cards represent the journey we pass through in life, and it is no surprise to find that the Moon appears not once, but twice. And in true lunar fashion, one appearance is a little less obvious than the other!

The goddess of the Moon appears as The Priestess, while another of the major arcana cards is simply called The Moon. This card is often interpreted as the unconscious mind, because of the influence of the Moon over the waters, and our own inner, psychic tides. The two cards are, in a way, opposed: the lunar goddess of intuitive wisdom is a positive aspect of the female principle, while the Moon card is often seen as one of dreamy delusion, the negative aspect. The two draw together, poetically and evocatively, the many different aspects of the Moon.

There are many versions of the tarot cards. Here we look at the images in the traditional Ryder-Haggard deck, which depicts the subtleties so beautifully.

The High Priestess is shown as a wise woman in a robe which flows like water towards the ground, spreading over a crescent Moon. She sits between two pillars, representing good and evil powers, and behind her a tapestry depicts rich fruits on the point of bursting from their pods. She holds a scroll of mystical teaching, and her head is crowned with two horn-like crescents and a circle, the full moon.

The card is number two of the major arcana, indicating balance, but also dualism. It suggests the polarity out of which all things are created, but also the fact that man is separated from the world around him, he is no longer at one with existence. It is the reflected light of the Moon, not the Sun's direct light.

The High Priestess is a distillation of the many Moon Goddesses down the ages. The image of the crowned head is like depictions of Isis. Her positive aspect is the *anima* – the feminine element in man – Diana, and Sophia, the Gnostic goddess of wisdom, lady of light. Her dark aspect appears when the power of the feminine principle is abused or ignored. Then she assumes the form of Hecate, Queen of the Dark Moon, and Lilith, Ruler of Demons.

There is the mystery of life itself here – which we feel when we gaze at the Moon. She sits between the pillars of negative and positive power, absorbing and unifying them, so that life may be created. It is the task of The Fool – who has appeared as the number one tarot card – to learn her secret as he begins his journey. She symbolizes, too, the connection between the conscious and the unconscious minds, she sits at the doorway into the psyche.

When this card is drawn in a tarot reading, it signifies that things previously hidden will be revealed, bringing insight and strength and the ability to solve problems. It may suggest the presence of a wise and intuitive woman, or may indicate your own inner wise woman.

THE FOOL.

THE MOON.

Creativity is also shown, just as the Moon mysteriously changes and recreates itself. If the card is reversed (upside down), it may suggest that emotional instability needs to be addressed, as this may be swaying judgment.

The card of The Moon represents the end of another stage of the quest.

From a fathomless pool a crayfish climbs out on to a long and winding path. On either side of it a fierce-looking dog and wolf bay at the full moon, which seems to be drawing water towards it. Two pillars – like gateways to an unknown land – stand on either side.

The ancient belief that the Moon was the dwelling place of the dead is suggested. But of course the Moon also governs fertility and new life – thus the two pillars, of death and of new life. The card shows that the time has come to abandon logical thought: to proceed along the winding path, the non-rational light of the inner self is needed.

The card is quite a turning point; danger lurks, the crayfish emerging from the depths may have his life sapped by the Moon. The dogs, symbols of the realm of the dark Moon goddess Hecate, may threaten.

But if the path is followed unerringly, the seeker will win through and reach the light.

When drawn in a reading, the card indicates that some kind of crisis of faith is underway, but that progress will be made if one's own intuition is relied upon. When reversed, it indicates that the nerve to do this – to go beyond the rational – is failing, and needs to be boosted.

The Moon, then, is not a comfortable card to draw, but it is a rich one. It shows that the final stage of some trial must be endured – a dark night of the soul in which all appears shifting and illusory. But faith and intuition, the Moon's cool light, will carry the inquirer through.

Chapter 3

Moonshine Rules

Living by the Moon

T he Moon has such a strong influence on the Earth that twice a day it makes the waters of the oceans ebb and flow. If it can perform such a feat, may it not also affect other parts of life on Earth? Plants, the weather, animals, man's physical being – all have been thought to be subject to lunar influences. Way back in the fourth century, Aristotle reckoned that oysters and sea-urchins were meatier during the full moon – and twentieth century scientific investigation has proved him right. Other old-wives' tales are proving to have more than a grain of truth in them.

> *"If you with flowers would stick the pregnant earth*
> *Mark well the moon proportions to their birth,*
> *For earth the silent midnight green obeys*
> *And waits her course, who clad in silver rays*
> *The eternal round of time and seasons guides."*
> *(The English Gardener)*

The idea of planting crops, flowers and trees according to the Moon's phases does sound strange to us today. How could the Moon make any difference to growth? But there has long been a belief that the Moon plays an important part in plant development, and modern research is now supporting it. In general, it is thought best to plant a couple of days before the full moon, when the Moon is on the increase. This seems logical. But the science of moon-planting turns out to be a bit more complex than that.

We all know of the lunar influence that causes the Earth's waters to ebb and flow twice a day, but does this powerful force also effect other things? Its effect on sea-life has been charted since ancient times and is now being scientifically verified.

A time to sow, a time to reap – all aspects of farming have been ascribed their proper lunar timing for the growth of a good crop.

"Picking medicinal herbs must be done when the Moon is in the sign of the Virgin, and not when Jupiter is in the ascendant, for then the herb loses its virtue." (Paracelsus, sixteenth century)

A certain Anthony Askham reckoned he had found a plant called lunary whose growth related directly to the Moon: "The leves of this herbe be rounde and blew and they have the mark of the Moone in the myddes . . . this groweth in the new Moone without leve to the end of fyftene dayes and after fyftene dayes it looseth every day a leve as the Moone waneth."[94]

That plant may only grow in the moonstruck imagination, but the effects of the Moon on plants were taken seriously. An East Anglian farmer wrote in 1562:

> "Sow peason and beanes in the wane of the Moone,
> Who soweth them sooner, he soweth too soone,
> That they with the planet may rest and arise,
> And flourish with bearing most plentiful wise."

A certain John Woolridge, around the same time, added: "The seeds from which you expect to have double Flowers, must be sown at the Full of the Moon or in two or three days after. It hath been long observed that the Moon hath great influence over plants . . .. and if it hath any such influence, then surely it is in the doubling of Flowers."

Even the renowned herbalist Nicholas Culpepper wrote in his famous *Herbal* that the stars influence man, the illnesses he is prone to, and the herbs which can cure him. "Consider what planet causeth the disease . . . by what planet the afflicted part of the body is governed. You may oppose diseases by Herbs of the planet . . . every plant cures his own disease; as the Sun and the Moon by their Herbs cure the Eyes."

Medieval herbalists believed that flowers produced twice as many blooms if they were sown around the time of the Full Moon.

It certainly seems logical to believe, as many herbalists do, that a plant's vital energies flow upwards during the waxing of the Moon, meaning that this is the time to gather the stalks, leaves and flowers, and that the energies are in the roots when the Moon wanes. Farmers thought that crops should be planted at the waxing moon, but that destructive work, such as killing pests and weeding, should be at the waning. There were particular recommendations for various plants: potatoes should be planted in the dark of the Moon, if possible on Good Friday; peas should be sown in the light of the Moon; root crops are best between the first quarter and full moon; leafy plants should be planted at the turn of the lunar month when the Moon begins to wane. And the Amish of Pennsylvania even plant fence posts according to Moon phases.

It was believed that pasture land could be greatly improved if the livestock only grazed it during the first and second quarters of the Moon. The animals should be removed after the full moon, during the third or fourth quarters. A book of 1600, *The Garden of Eden*, urges cultivation of flowers according to Moon phases. It claims that single flowers like the gillyflower and tulip can be made to double if they are transplanted three days after the full moon. A seventeenth-century book, *The Expert Gardener*, says that violets, rosemary and lavender should be sown when the Moon is new, but wallflowers when it is old. Here are a few more old-wives' tales concerning plants and the Moon:

As well as being planted according to the position of the Moon, flowers and other plants were gathered by herbalists at certain Moon times, when their vitality was believed to be at its height.

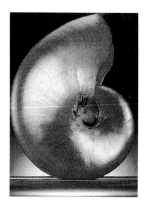

According to the English herbalist Culpepper, moonwort has the power to pull the shoe from a horse that steps on it.

"Don't plant seed too soon,
Consult the Moon."

A full moon at Christmas means that next year's harvest will be poor. According to occult tradition, herbs should only be planted while the Moon is in one of the moist, fertile signs – these are Cancer, Pisces, Taurus and Capricorn.

Trees, needing to develop strong roots, should be planted during the waning Moon, ideally after the last quarter but before the New Moon.

Flowers valued for their fragrance, such as lavender, should be planted during the Moon's first quarter, ideally in the sign of Libra, to achieve best fragrance.

If an abundance of flowers is the main aim, plant while the Moon is in Cancer, Scorpio or Pisces.

Flowers, herbs and trees for timber were all thought to be best gathered when the Moon is declining, because they yield much better, being weaker during the Moon's wane.

Trees should be pruned, however, during the Moon's increase, if they are to produce abundant fruit.

There is still a widely-held belief that timber should only be cut when the Moon is waning, because at this time the sap moves down and decreases, making the timber drier and easier to cut. In other words, trees too are subject to tidal pull. Some carpenters refuse to use wood cut during the waxing moon, saying that the excess moisture in it will make it warp. Specific rules for trees have evolved over the years:

Hard woods, such as oak and chestnut, should be cut before noon after the full moon. White woods, such as pine and maple, should be cut again before noon, but between the new moon and full moon, in the sign of Virgo.

If you want to prevent a tree or shrub from growing too large, plant or prune it during the dark moon, when it is in Cancer. Trees which you wish to grow fast should be pruned in the Moon's first quarter.

Never cut timber during the light of the Moon.

An Arkansas farmer experimented with grafting fruit trees according to Moon phases. He concluded that the best time for this was when the Moon

was between new and full, between the first and second quarters. He further recommended that the process be carried out when the Moon was in Cancer, this being "the most fruitful, movable, watery and feminine of signs".[95]

Some flowers are said to have particular connections with the Moon. The poppy is considered to be a flower of the moon, perhaps because of its association with death. The rose is associated with Diana. It was originally a Grecian queen, so beautiful that Diana was jealous.

Diana's brother, the Sun God, scorched the queen until she shrank into a flower, and her three lovers who had pursued her to Diana's temple were also punished, turned into a butterfly, a worm and a drone.[96]

If white flowers are offered to the goddess of the Moon on the night that the new moon can first be seen, a fortunate month will follow. White flowers are also believed to come under the Moon's influence.

They were thought to be inhabited by Moon spirits, who appeared at full moon, especially in July, August and September. Jasmine, too, is a flower of the Moon, and of the mysteries of the night. Its oil is used to attract love, and the scent of jasmine helps to bring sleep and to aid meditation. Perhaps all this advice about planting should not be dismissed as old wives' tales – there is no doubt that animals are affected by the Moon's phases. The movement of fish in the sea, the spawning of various marine animals such as crabs, mussels, oysters and sea urchins – all these are done according to lunar periods.

Even large animals have been observed to have an increased mating urge during the waxing of the Moon. Many farmers in the past have managed their stock according to the Moon: pigs were slaughtered when the Moon was waxing, because the bacon was considered richer and fatter than during the wane. Sheep were sheared at this time too, when the wool was found to be richer and thicker. Farmers reckoned that animals born during a waning moon would be sickly, and they would never geld their stock at this time, to prevent them from sickening.

While mortals sleep, and dream their dreams beneath the Moon, its subtle power is influencing the growth of life on our planet.

Nothing to do with the Moon is ever absolutely straightforward! There are even Moon-rules about hatching eggs, but you may need to be an astrologer to work it all out.

Eggs were set so that they would hatch during the light of the Moon, when it is in Cancer, Scorpio or Pisces. The chicks hatched in the new moon are supposedly healthier and grow faster than those hatched under the old moon; and those hatched when the Moon is in a fruitful sign of the zodiac will be good layers. A colt should be weaned when the Moon is in Capricorn, Aquarius or Pisces. Animals should also be slaughtered according to Moon signs. Meat is said to have a better flavor, be more tender and keep longer if the animal is killed three days after a full moon, ideally in the sign of Leo.

Is there any modern evidence that the Moon really does affect animal life? Plenty! It even begins to seem ridiculous that we could have doubted it for so long. Lyall Watson, in the now classic *Supernature*, can take some of the credit for restoring our belief in the wonders and mysteries of nature. He gives many examples of the Moon's very real influence on animal life:

A small flat-worm – *convoluta* – which lives in the sand, and has to rise to the surface when the tide goes out to receive sunlight. When kept in a tank in a laboratory, it continues to follow the tidal rhythms. This has since been found to be true of any normally tidal animal.

Oysters which were taken 1,000 miles from their habitat, to a laboratory inland, soon adjusted their behavior (feeding at high tide, closing at low tide), to what the tides would have been at their new location, if it had been by the sea. This was true even if the oysters were kept in dark containers – in other words, they were definitely reacting to the Moon's passage.

A small silver fish, the grunion, makes use of the difference between spring and neap tides. Soon after the full moon, between March and August, just before the spring tide starts receding it lays its eggs on the wet sand by some amazing feats of acrobatics between shore and waves. The eggs, just above

the high tide line, stay undisturbed until the next spring tide can reach them, by which time the larvae are ready to hatch when touched by the water.

The sooty tern of Ascension Island breeds only at the tenth Full Moon.

Mating takes place on this night (which is why the birds are known locally as "wide-awakes"). They then fly away, and only return to the island at the time of the next tenth full moon.

Hamsters studied over four years were found to follow a lunar rhythm.

A peak of activity occurred four days after the full moon, and there were also consistently high levels for a few days after the New Moon.

When it comes to the weather, we are back in the realms of ancient folklore. But are things really so different today?

Omens of coming weather have always been seen in the Moon, because it governs the waters. In India it was thought that the rain came from the Moon, and rainfall has always been associated with phases or particular aspects of the Moon.

If two Moons occur in the same calendar month, especially May, floods and other calamities may be expected.

Rings around the Moon are thought to herald storms.

"A ring around the Moon,
Rain comes soon."
"If the Moon shows a silver shield,
Be not afraid to reap your field.
But if she rises haloed round
Soon we'll walk on deluged ground."
"When the Moon lies on her back,
Then the sou'west wind will crack,
When she rises up and nods,
Chill nor'easters dry the sods."

"If the lower horn of the Moon is dusky, it will rain before the Full Moon. If the horns of the Moon are sharp on the third day, the whole month will be fine. If the upper horn of the Moon is dusky at setting, it will rain during the wane of the Moon. If the center is dusky, it will rain at Full Moon."[97]

If thunder happens when the Moon is changing phase in spring, the weather will be mild and moist, promising good crops.

If the full moon happens at the equinox, there will be violent storms, but a very dry spring will follow.

However many days the Moon is old on Michaelmas Day (September 29) is how many floods there will be afterwards.

If the Moon casts no shadows by the time it is four days old, prepare for bad weather.

"An old moon in a mist is worth gold in a kist,
But a new moon's mist will never lack thirst." [98]

"When round the Moon there is a brugh (halo)
The weather will be cold and rough."
The dew is said to fall heaviest on moonlit nights.
Clear Moon, frost soon.
"Pale Moon does rain, red Moon does blow
White Moon does neither rain nor snow."

If the Moon changes its phase close to midnight, good weather will follow for the next seven days. If the change happens nearer noon, the weather will be more changeable. A new moon on a Saturday precedes foul weather. A New Moon on a Monday – "Moon Day" – is a lucky sign, and brings good weather. But if the Moon appears on a Sunday, there will be floods before the month is out.

Lyall Watson tells the story of two quite independent studies which were done in the early 1960s, in Australia and the USA respectively, "both coming to the same conclusions but reluctant to publish their findings for fear of ridicule. Only when each discovered the existence of the other and learned that their findings had been confirmed, did they go to press – together, for mutual support, in the same magazine.[99]

The American study was of weather data covering forty-nine years, and found that there was increased rainfall in the first and third weeks of the synodical month – i.e., on the days after the full and the new Moon. The Australians, looking at data over twenty-five years, found the same patterns. Watson suggests that the cause of this could be the greater precipitation of meteor dust at the times of full and new moons, causing greater condensation of water vapor in clouds, and hence rain.

What about man, does the Moon also affect his physical body? The Moon has been compared with the brain. Aristotle saw them both as cold, moist and insensitive! He reckoned the brain became moister and fuller under a waxing moon than under a waning one. Not so far removed from the ever popular

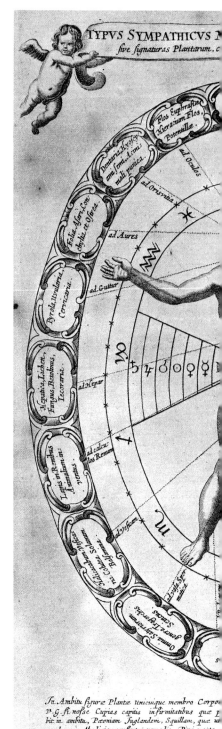

TYPVS SYMPATHICVS M
five fignaturas Plantarum, c

In Ambitu figuræ Plantæ unicuique membro Corpo...
v. g. si nosse Cupias capitis infirmitatibus quæ ...
bit in ambitu, Pæoniam Juglandem, Squillam, quæ ...
morbos à Medicis confentur remedia. Pari pacto

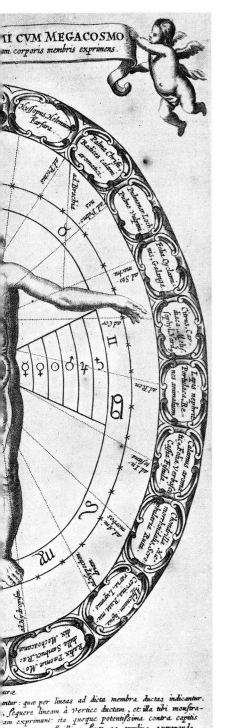

belief that the Moon can affect people's minds, leading to "lunacy" and becoming "moon-struck"'. One belief commonly held, particularly among people living in coastal regions, was that there was a much higher birth rate when the Moon is full over the local meridian, because it affects uterine contractions. Many astrologers have said that the Moon is connected with the blood flow.

In ancient India surgeons would only operate when the Moon was waning, to prevent scarring.

Let us take a look at the evidence of connections betweeen man and the Moon.

According to a 1940s study of tuberculosis, deaths from the illness are at their height seven days before full moon. It suggests this is linked to a lunar cycle in the PH content (the ratio of acid to alkali) of the blood.

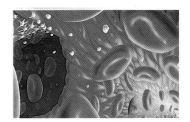

A similar correlation has been found with pneumonia, the amount of uric acid in the blood, and the time of death.

The incidence of hemorrhages after throat operations was found to be eighty-two per cent greater during the second quarter of the Moon, in a Florida study. The problem cases were clustered around the full moon, with the minimum at new moon.

Similar results were found in analysis of bleeding peptic ulcers. An Irish study found that recovery from fractures of the head of the femur varied according to the lunar phase.

One spin-off from the race to the Moon was that the huge amount of research done in the 1960s revealed many of the Moon's effects on the Earth. Studies published in the *Journal of Geophysical Research* showed monthly lunar patterns in: the number of meteors striking the Earth; the number of hail-like ice lumps which form in high-altitude clouds; the amount of ozone in the atmosphere; the times of heavy rainfall in the USA and New Zealand; the Earth's magnetic field was also found to follow lunar cycles. Perhaps it is scientists who are now more convinced than anyone of its influence!

Left: The belief that the planets govern the body may have been long ago discarded as superstitious nonsense, but modern medicine is now finding some curious correlations.

Chapter 4

The Moon in Literature

The following are a few exerpts showing writers' love of the moon.

"Late, late yestreen I saw the new moon,
With the auld moon in her arm." [100]

"With how sad steps, O moon, thou climb'st the skies!
How silently, and with how wan a face!" [101]

"I'd rather be a dog and bay the moon
Than such a Roman." [102]

"He opens his hand to grasp the moon in heaven;
He plunges into the sea to seize leviathan."
(Chinese Proverb)

"But, oh! I talk of things impossible,
And cast beyond the moon." [103]

"I was predisposed to notice in prison a strange kind of longing among the Japanese for the Moon; and out of this longing a kind of frustration and a kind of rage over the frustration which added to the power of this very deep longing. It was just like a tide in the sea being pulled up in their characters and this seemed to be most evident as the Moon swelled and became more and more full. And one became quite frightened of them . . .

"I've seen the phenomenon in Africa. I have heard lions roar at the Moon just as dogs bay at the Moon. I heard a lion one night start as the Moon came up at 8 o'clock and, so obsessed was he with it that he was still roaring at it in full daylight after it had set the next day. But this phenomenon I watched so closely and anxiously seemed to wane with the Moon, and a very pure, almost resolved kind of excitement and expectation built up within the Japanese. They were almost joyful and suddenly, when the New Moon appeared in the sky, there was a feeling of catharsis, of having been cleansed, among them, as if somehow they had done the journey themselves. You know, they had grown with the Moon, they had waxed and they had gone back into the darkness and were emerging into the light again . . . it was so mysterious." [104]

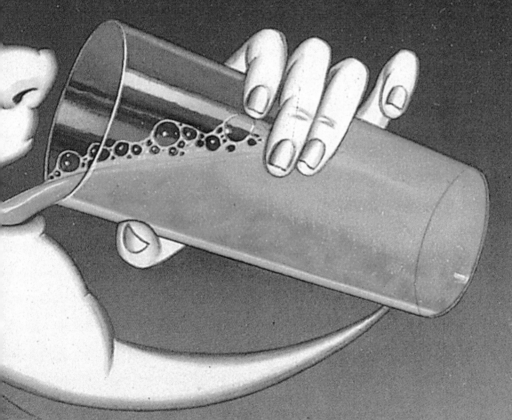

"Most of the really bad excesses (in the Japanese POW camp) . . . occurred about the time the Moon started waning, when it started going back into the darkness. You see, this tide of longing was drawn absolutely to the full, and then suddenly what was there but the night again? . . . I often wondered if there were a rearrangement, a reversal of the poles in the Japanese spirit, with the Moon a sort of implied masculine . . . for the Germans too the Moon is masculine and the Sun is feminine and these things are signposts of the spirit; they and we should perhaps not think, so much as wonder, aloud about such things."[105]

The Moon occupies a special place in the literature of many countries. It is the stuff that dreams are made of, the silver disc beneath which lovers make their vows, but also a harbinger of change and heartache.

"There can therefore be but two Full Moons in the year which rise during a week almost at the same time as the Sun sets; the former, occurring in September, is called the Harvest Moon; and the latter, in the month of October, being in a similar predicament, is termed the Hunter's Moon."[106]

"The sun set and uprose the yellow Moon:
The devil's in the moon for mischief; they
Who called her chaste, methinks, began too soon
Their nomenclature; there is not a day
The longest, not the twenty-first of June,
Sees half the business in a wicked way
On which three single hours of moonshine smile."
(Don Juan)

"Soon as the evening shades prevail
The moon takes up the wondrous tale;
And nightly, to the listening earth,
Repeats the story of her birth."
(Addison, 1712)

"Or the coy moon, when in the waviness
Of whitest clouds she does her beauty dress
And staidly paces higher up, and higher,
Like a sweet nun in holiday attire." [107]

"The man who has seen the rising moon break out of the clouds at midnight has been present like an archangel at the creation of light and of the world."[108]

"Yon rising moon that looks for us again –
How oft hereafter will she wax and wane;
How oft hereafter rising look for us
Through this same garden – and for one in vain!" [109]

"Lo, the moon ascending,
Up from the east the silvery round moon,
Beautiful over the house-tops, ghastly, phantom moon,
Immense and silent moon." [110]

"How sweet the moonlight sleeps upon this bank!
Here will we sit and let the sounds of music
Creep in our ears." [111]

"When the moon shone, we did not see the candle;
So doth the greater glory dim the less." [112]

"It glimmers on the forest tips
And through the dewy foliage drips In little rivulets of light." [113]

"On moonlight nights the Recording Angel uses shorthand."
(Anon.)

"Girls and boys come out to play,
The Moon doth shine as bright as day.
Come with a whoop, come with a call,
Come with a goodwill or not at all."

"There was an old woman tossed up in a basket
Seventy times as high as the moon.
Where was she going? I couldn't but ask it,
For in her hand she carried a broom.
'Old woman', quoth I 'Oh whither so high?
'To sweep the cobwebs off the sky.'"

"Au clair de la lune, mon ami Pierrot,
Ma chandelle est morte, je n'ai plus de feu . . ."

"Hey diddle diddle
The cat and the fiddle
The cow jumped over the moon;
The little dog laughed
To see such sport
And the dish ran away with the spoon."

This last rhyme was first printed around 1785, and ever since there have been many interpretations of its meaning. They range from pagan worship rites to reference to star constellations (Taurus, Canis minor, etc.) to allusions to Queen Elizabeth and Lady Katherine Gray. And many more! It may have come from the old game of cat (trap-ball) and fiddle (music), provided in some old inns. But as one observer commented, "I prefer to think that it commemorates the athletic lunacy to which the strange conspiracy of the cat and the fiddle incited the cow."

Above: Illustrations for "Hey diddle diddle" by Edward Lear, the "nonsense" poet. It has been suggested that the rhyme originally had hidden meaning.

Moon-viewing (*tsukimi*) is a common theme in Japanese poetry, and some of the most beautiful traditional poems use the Moon as a symbol of desire and love. It is as popular in Japanese poems as the themes of Snow-Viewing and Cherry Blossom-Viewing.

This famous haiku by the ninth-century poet Narihira uses the Moon to conjure up an experience of ill-fated love.

"Tsuki ya aranu
Haru ya mukashi no
Haru naranu
Waga mi hitotsu wa
Moto no mi ni shite"

The translation of which is:

"Moon, are you not the same?
Spring, can it be that you are not
The spring of old,
And I myself the only thing
Remaining, as it used to be?"

Another well-known poem, by the ninth-century female poet, Komachi, demonstrates the very practical importance of the Moon to lovers of old.

"Hito ni awan
Tsuki no naki ni wa
Omoiokite
Mane hashiribi ni
Kokoro yake ori"

Meaning:

"On such a night,
When no Moon gives us the chance to meet,
I wake my passion blazing,
In my breast an uncontrollable fire
That utterly consumes my heart."

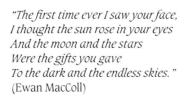

"The first time ever I saw your face,
I thought the sun rose in your eyes
And the moon and the stars
Were the gifts you gave
To the dark and the endless skies."
(Ewan MacColl)

"But soft! What light through yonder window breaks?
It is the east, and Juliet is the sun!
Arise, fair sun, and kill the curious moon,
Who is already sick and pale with grief,
That thou her maid art far more fair than she:
Be not her maid, since she is envious;
Her vestal livery is but sick and green,
And none but fools do wear it." [114]
Romeo

"Lady, by yonder blessed moon I swear,
That tips with silver all these fruit-tree tops,

Juliet: O, swear not by the moon, the inconstant moon,
That monthly changes in her circled orb,
Lest that thy love prove likewise variable." [115]

Moonshine: This lantern doth the horned moon present;
Myself the man i' the moon do seem to be.

Theseus: This is the greatest error of all the rest:
the man should be put into the lantern.
How is it else the man in' the moon? . . .

Moonshine: All that I have to say,
is to tell you that the lantern is the moon;
I, the man in the moon;
this thorn-bush, my thorn-bush;
and this dog, my dog.

Demetrius: Why, all these should be in the lantern;
for all these are in the moon." [116]

". . . now glow'd the Firmament With living Saphirs:
Hesperus that led
The starrie Host, rode brightest, till the
Moon Rising in clouded Majestie, at length
Apparent Queen unvaild her peerless light,
And ore the dark her Silver Mantle threw." [117]

"Thus at their shadie Lodge arriv'd, both stood,
Both turn'd, and under op'n Skie ador'd
The God that made both Skie, Air, Earth and Heav'n
Which they beheld, the Moons resplendent Globe And starrie Pole.
Thou also mad'st the Night, Maker Omnipotent, and thou the Day." [118]

The Full Moon conjures up ideas of the little folk:

"Up the airy mountain,
Down the rushy glen,
We daren't go a hunting
For fear of little men;

    Wee folk, good folk,
    Trooping all together;
    Green jacket, red cap,
    And white owl's feather!" [119]

A Dancing Butterfly

The respectable, Christian Victorians probably had no idea that this poem echoed more pagan beliefs of the Moon as the home of the dead.

"I heard the dogs howl in the moonlight night,
And I went to the window to see the sight;
All the dead that I ever knew
Going one by one and two by two.
    On they pass'd, and on they pass'd;
    Townsfellows all from first to last;
    Born in the moonlight of the lane
    And quench'd in the heavy shadow again.
        On, on, a moving bridge they made
        Across the moon-stream, from shade to shade
        Young and old, women and men;
        Many long-forgot, but remembered then.
            And first there came a bitter laughter;
            And a sound of tears a moment after;
            And then a music so lofty and gay,
            That every morning, day by day,
            I strive to recall it if I may."
        (Allingham, A "Dream")

"Deep on the convent-roof the snows
Are sparkling to the moon:
My breath to heaven like vapour goes:
May my soul follow soon!
All heaven bursts her starry floors,
And strows the light below,
And deepens on and up! the gates
Roll back, and far within
For me the Heavenly Bridegroom waits,
To make me pure of sin." [120]

"Then, ere the silver sickle of that month
Became her golden shield, I stole from court
With Cyril and with Florian, unperceived,
Cat-footed through the town and half in dread
To hear my father's clamour at our backs." [121]

"The minstrels played their Christmas tune
To-night beneath my cottage eaves;
While smitten by a lofty moon,
The encircling laurels, thick with leaves,
Gave back a rich and dazzling sheen
That overpowered their natural green." [122]

*"Tender is the night
And haply the Queen-Moon is on her throne,
Cluster'd around by all her starry Fays;
But here there is no light,
Save what from heaven is with the breezes blown
Through verdurous glooms and winding mossy ways."* [123]

*"This is the light of the mind, cold and planetary.
The trees of the mind are black.
The light is blue.
The grasses unload their griefs on my feet as if I were God,
Pricking my ankles and murmuring of their humility.
Fumey, spiritous mists inhabit this place
Separated from my house by a row of headstones.
I simply cannot see where there is to get to.
The moon is no door.
It is a face in its own right,
White as a knuckle and terribly upset.
It drags the sea after it like a dark slime; it is quiet
With the O-gape of complete despair.*

*I live here.
Twice on Sunday, the bells startle the sky –
Eight great tongues affirming the Resurrection.
At the end, they soberly bong out their names.
The yew tree points up. It has a Gothic shape.
The eyes lift after it and find the moon.
The moon is my mother. She is not sweet like Mary.
Her blue garments unloose small bats and owls.
How I would like to believe in tenderness
The face of the effigy, gentled by candles,
Bending, on one in particular, its mild eyes.
I have fallen a long way. Clouds are flowering
Blue and mystical over the face of the stars.
Inside the church, the saints will be all blue,
Floating on their delicate feet over the cold pews,
Their hands and faces stiff with holiness.
The moon sees nothing of this.
She is bald and wild.
And the message of the yew trees is blackness –
blackness and silence.* [124]

"Though the night was made for loving,
And the day returns too soon,
Yet we'll go no more a-roving
By the light of the Moon." [125]

"The stars were dim, and thick the night
The steersman's face by his lamp gleamed white;
From the sails the dew did drip –
Till clomb above the eastern bar
The horned Moon, with one bright star
Within the nether tip.
One after one, by the star-dogged Moon,
Too quick for groan or sigh,
Each turned his face with a ghastly pang,
And cursed me with his eye." [126]

"'Twas night, calm night, the Moon was high;
The dead men stood together.
All stood together on the deck
For a charnel-dungeon fitter:
All fixed on me their stony eyes,
That in the Moon did glitter." [127]

"They make them beleeve, according to the Proverbe, that gloe wormes are lanternes, and that the Moone is made of greene Cheese." [128]

"Is the moon tired?
She looks so pale
Within her misty veil;
She scales the sky from east to west,
And takes no rest.
Before the coming of the night
The moon shows papery white;
Before the dawning of the day
She fades away." [129]

"Crazed through with much child-bearing
The moon is staggering in the sky;
Moon-struck by the despairing
Glances of her wandering eye
We grope, and grope in vain,
For children born of her pain." [130]

"What is there in thee, Moon!
   that thou shouldst move
   My heart so potently?
   When yet a child I oft have dried my tears
   when thou hast smil'd.
   Thou seem'dst my sister: hand in hand we went
   From eve to morn across the firmament." [131]

"Lady Moon, Lady Moon, where are you roving?
Over the sea. Lady Moon, Lady Moon, whom are you loving?
All that love me. Are you not tired of rolling, and never
Resting to sleep? Why look so pale, and so sad, as for ever
Wishing to weep? Ask me not this, little child! if you love me;
You are too bold; I must obey my dear
Father above me, And do as I'm told.
Lady Moon, Lady Moon, where are you roving?
Over the sea. Lady Moon, Lady Moon, whom are you loving?
All that love me." [132]

"Softly, silently, now the moon
Walks the night in her silver shoon;
This way, and that, she peers, and sees
Silver fruit upon silver trees;
One by one the casements catch
Her beams beneath the silvery thatch;
Couched in his kennel, like a log,
With paws of silver sleeps the dog;
From their shadowy cote the white breasts peep
Of doves in a silver-feathered sleep;
A harvest mouse goes scampering by,
With silver claws, and silver eye;
And moveless fish in the water gleam,
By silver reeds in a silver stream." [133]

"Is there anybody there? said the Traveller,
Knocking on the moonlit door
But only a host of phantom listeners
That dwelt in the lone house then
Stood listening in the quiet of the moonlight
To that voice from the world of men:
Stood thronging the faint moonbeams on the dark stair,
That goes down to the empty hall." [134]

"Tell me what you feel in your room when the Full Moon is shining in upon
you and your lamp is dying out, and I will tell you how old you are, and I shall
know if you are happy." [135]

"Emotional expression is infinitely rich and varied of form; the Moonlight causes a Yankee butcher to say to his wife:

'It's such a beautiful night I can't lie still another minute; I must go out and do some slaughtering.'"[136]

"The man in the Moon was caught in a trap,
For stealing the thorns from another man's gap,
If he had gone by, and let the thorns lie,
He'd never been man in the Moon so high."

"In some few seconds the watchman had accomplished the 240,000 miles to the moon . . . there stood a city, which we can only imperfectly imagine if we beat up the white of an egg in a glass of water. The material was quite as soft, and formed just such towers and domes, transparent and wavering in the thin air. Our Earth hung above his head like a fiery ball.

"The Moonites were disputing about our earth, and doubted that it could be inhabited; for certainly, they said, the air must be too thick there to allow a rational creature to breathe at all. They considered the moon alone to be inhabited, and that it was the heart of the whole planetary system. What strange ideas men, or rather Moonites, have!"[137]

"Softly, silently, now the moon
Walks the night in her silver shoon;
This way, and that, she peers, and sees
Silver fruit upon silver trees."

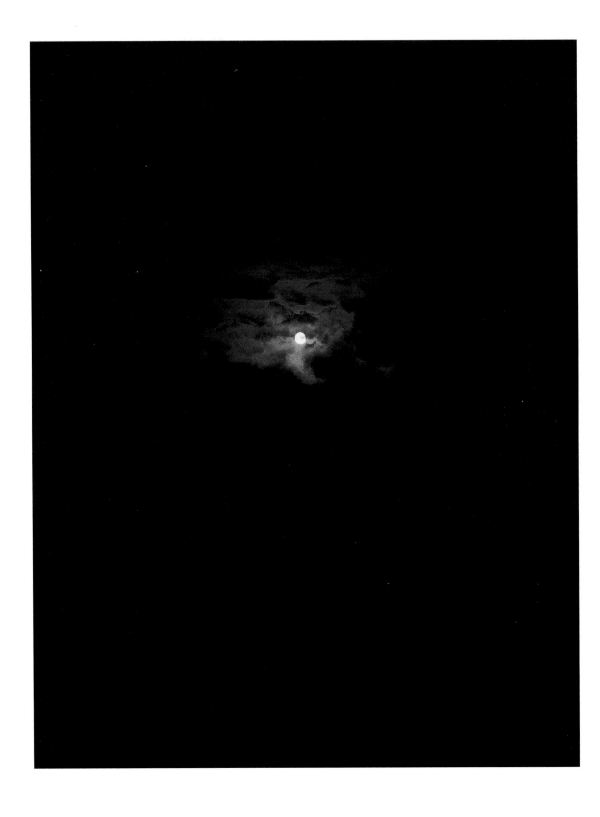

Chapter 5

Moon-Food

In this chapter we give you, for your delectation, a few ways of feasting with the Moon.

China Moon

The Moon has always held a special place in China, and the mid-August moon – comparable to the harvest moon of the West – is considered the most beautiful of the Moon's aspects.

This is the time of the Moon Festival, when people set out in the evenings to view the Moon, to read and write poetry – and to eat Mooncakes. These are perfectly round cakes made of wheat flour and brown sugar and stuffed with sweets.

In the north there are only two kinds of filling – white sugar paste or brown date paste. In the south the fillings may be made from ham, dates or preserved apricots, walnuts, lard and watermelon seeds.

Many festivals are woven around the Moon in both Japan and China. The Chinese have even invented special Mooncakes in its honor.

Mooncakes

This recipe for Mooncakes comes from Shanghai. Ideally, they should be made in Chinese Moon Cake molds, as this imprints the traditional chrysanthemum pattern and Chinese characters on them. However, you can easily mold them yourself; they should be about three inches (seven centimeters) in diameter. You can try drawing your own designs on them before baking – invent your own "traditional patterns"!

4 cups flour
4 tbsp. brown sugar
half tsp. salt
4oz (110g) margarine
1 egg
1 tsp. sesame oil

For the filling:
2 tbsp. peanuts
2 tbsp. sesame seeds
2 tbsp. walnuts or pine nuts
2 tbsp. chestnuts, boiled until tender, or blanched almonds
2 tbsp. sultanas or other dried fruit, chopped
2 tbsp. chopped dried apricots
4 tbsp. brown sugar
2 tbsp. margarine
2 tbsp. rice flour or poppy seeds

Preheat oven to 400f/200c/Gas Mark 6. Yield about 16 cakes. Sift the flour, sugar and salt together. Chop the margarine into pieces and rub into the flour until crumbs form. Add enough hot water (about half a cup) to make a pastry dough. Cover with a cloth. Roast the peanuts in a hot pan for two minutes. Add the sesame seeds, then put a lid on to stop them from jumping out of the pan. Roast for a further two minutes. Put the peanuts and seeds in a food processor or blender and grind with the other nuts. Add to the rest of the filling ingredients and mix together. Roll out the pastry on a floured board. Cut rounds with a pastry cutter to fill the mold – if you have one – or make little pie cases. Rub the mold with margarine and spread pastry over the bottom and sides of the mold. Put in a tablespoon of filling. Press down gently. Wet the edges of the pastry and cover with another round to make a lid. Seal together, and remove from the mold if you are using one. Put all the cakes on a greased baking sheet. Beat the eggs and sesame oil together and brush each cake with this mixture. Bake about thirty minutes until the cakes are golden brown.

This makes about sixteen cakes. (Jack Santa Maria, *Chinese Vegetarian Cookery*, Rider and Co., London, 1983) Along with the Mooncakes, ripe melons, green soybeans and fruits would also be laid out in the garden, in thanksgiving.

Soybeans are set out in the Chinese garden as part of a lunar thanksgiving festival.

Moon-Shine

This recipe for Moon-Shine comes from an eighteenth-century English recipe book, *The Art of Cookery Made Plain and Easy*, by Mrs Glasse. The recipe is not entirely plain and easy, but quite spectacular. We present it as a curiosity, but you could substitue the elaborate meat recipe with your own favourite paté or vegetarian mixture. It is made in a large tin mold in the shape of a half-Moon, and in one large and two or three smaller star-shaped tins.

"Boil two calves feet in a gallon of water till it comes to a quart, then strain it off, and, when cold, skim off all the fat, take half the jelly, and sweeten it with sugar to your palate, beat up the whites of four eggs. Stir all together over a slow fire till it boils, then run it through a flannel bag till clear, put it in a clean saucepan, and take an ounce of sweet almonds blanched and beat very fine in a marble mortar, with two teaspoonfuls of rosewater, and two of orange-flower water; then strain it through a coarse cloth, mix it with the jelly, stir in four large spoonfuls of thick cream, stir it all together till it boils, then have ready the dish you intend it for, lay the tin in the shape of a half-moon in the middle, then pour in the above blanc-manger into the dish, and, when it is quite cold, take out the tin things, and mix the other half of the jelly with half a pint of good white wine, and the juice of two or three lemons, with loaf sugar enough to make it sweet, and the whites of eight eggs, beat fine; stir it all together over a low fire till it boils, then run it through a flannel bag, till it is quite clear, in a china basin, and very carefully fill up the places where you took the tin out; let it stand till cold, and send it to the table.

"Note, you may for change fill the dish with a fine thick almond custard; and, when it is cold, fill up the half-moon and stars with a clear jelly."

Moon Biscuits

Moon Biscuits are traditionally eaten with the wine that is blessed during the Drawing Down the Moon ritual (see page 101). They are made in the shape of the crescent moon, and the whole hazelnuts in them represent the Full Moon that is to come.

250g/9oz wholewheat flour
75g/3oz soft light brown sugar
175g/6oz butter or vegan marg
A good handful of hazelnuts

Heat the oven to 300f/150c/Gas Mark 2. Beat the butter or margarine with the sugar until blended. Add the flour, and mix together to form a dough. Knead on a floured surface. Gently work the whole hazelnuts through it, flatten out to a depth of about half an inch. If you have a moon-shaped pastry cutter, use this to form the biscuits. However, you may get more individual results if you cut the Moon shapes yourself with a small sharp knife. You can even add a few features, or Moon symbols, to the surface. Place the biscuits on a baking sheet, and put in the oven until light golden brown.

AU MAGOT
Rue St-James, 36, 38, 40 et 42, BORDEAUX

DERNIER QUARTIER

Chapter 6

Beyond the Moon

I t is more than twenty years since man first stepped on the Moon. That single historical event has radically changed the way we feel about the Moon. Some say that the great Moon Goddess has been violated by our visitation, that the poetry has gone out of the Moon, but is this really so? And will we ever set foot there again?

There are many things scientists would still dearly love to know about the Moon, in order to answer questions like: Was the Moon once able to support life, and if so of what kind? Did the Moon and the Earth develop separately or in tandem? Are there many more ores and minerals, unknown to us, still to be discovered there? Is there water somewhere on the Moon where man has not yet been, perhaps around the North Pole?

Above: Samples of moon rocks, such as this brought back by the Apollo 11 mission, have revealed some of the Moon's secrets.

Right: But if we take time to Moon-gaze, we may wonder if many mysteries still remain.

Manned lunar missions could happen again in our lifetime. The US national space policy is considering the future of space missions.

Among options under review are continued exploration of the solar system, manned missions to Mars, and the development of an outpost on the Moon. This last option might seem the least glamorous, as man has already been to the Moon, but it could provide the richest information. If it did happen, the astronauts would stop off at a space station before transferring to the lunar landing vehicle; they would stay on the Moon for two weeks at a time, would perhaps mine for oxygen, and would develop a Moon base in which thirty people could live for several months.

It is over twenty years since man last landed on the Moon, but if it happens again in our lifetime it might provide the resources for a space colony.

It has even been suggested that a giant telescope (possibly made from lunar glass) may one day be built on the dark side of the Moon – the natural place from which to observe deep space. The Moon could also be an ideal "laboratory" for discovering more about, for instance, superconductivity, and learning how to adapt to living on other planets.

Right now, though, the beauty and wonder of space exploration is taking a back seat to man's constant fight with man. The US has stated its "Star Wars" policy, in which the heavens become just another means of conflict and fighting over territory. It is also possible that another "cold war" could develop between the USA (greatly set back since the *Challenger* disaster), the USSR, the European Space Agency, Japan and China – all of them contenders in a potential new space race.

Another idea for the future is that the Moon might one day provide the resources for building and maintaining a space colony. This would be cheaper and easier than bringing everything from Earth, as the Moon has all that is needed in terms of iron, aluminium, titanium and magnesium. Compressed material would be loaded into a kind of sled, and "catapulted" into space along aluminium rails.

We still have our sights set on the Moon; scientists have many unanswered questions about its origins and its structure.

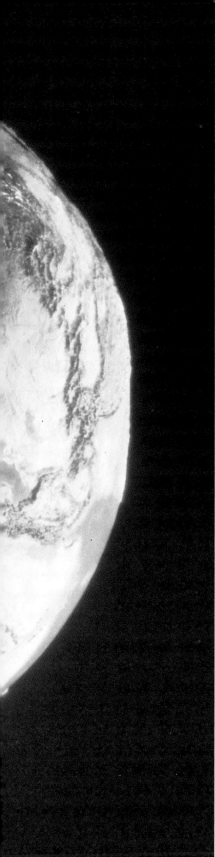

At least the Moon landing has finally squashed the idea that there is life on the Moon . . . Or has it? Not quite! It seems we give up such beliefs with difficulty. UFO-logy is the modern version of all those tall tales, and even since the Moon landing there have been stories of flying around the Moon in an alien spacecraft, and finding that the far side has an atmosphere, water, forests and settlements!

Perhaps Carl Jung should have the last word on this. Our desire to believe in life on other planets is, he says, a kind of projection. Life is harsh here, and becoming ever more threatened by global disaster. No wonder we look for some kind of saviors appearing from the heavens, or at least try to identify a hospitable civilization to which we may one day have to retreat. And if such a horror as an apocalypse happens, what of the Moon? The Book of Revelations uses it as a symbol:

"And . . . when he had opened the sixth seal . . . there was a great earthquake; and the Sun became black as sackcloth of hair, and the Moon became as blood; And the stars of heaven fell upon the Earth, even as a fig tree casteth her untimely figs, when she is shaken of a mighty wind."
(Revelations 6, 12–13)

But it also suggests that we will then not need the Moon:

"And the city had no need of the sun, neither of the Moon, to shine in it: for the glory of God did lighten it, and the Lamb is the light thereof." (Revelations, 21, 11–23)

So the Bible describes the ending of time, which is measured by the Sun and the Moon, and the coming of the kingdom of God when all opposites are united. Maybe such an apocalypse will not come to pass.

There is more and more interest in the Great Mother, the Moon Goddess – whatever we choose to call the creative, sensitive and deeply rooted part of ourselves.

The Moon Goddess has certainly not been forgotten: she has survived through the ages, hidden sometimes as the Moon itself is hidden when it is dark. But she is always there, always pulling on our innermost beings, touching us with her beauty, her fecundity, her mystery. Now she is waxing again, she is coming out for us all to be aware of as we recognize the need to develop those hidden "feminine" parts of us, so long neglected. Perhaps Isis, Diana, Artemis – whichever name you choose – will be our saving yet; for one of the most profound effects of the Moon-landing must be not what we learnt of that distant place, but what it showed us about our own home.

"The light of the Moon is borrowed, it is simply reflected. It is being reflected from the Earth too; you just have to be far away to see it. The astronauts could not believe the beauty of the Earth, looking from the Moon. The Moon was just ordinary. Not even grass grows there, no water, no beautiful mountains, no trees, no birds, no life . . . But looking from there to the Earth, the Earth looks so glorious, so beautiful."[138]

Part III
MOON SCIENCE

Chapter 1

The "Real" Moon

"He made his friends believe that the Moon is made of green cheese." [139]

The Moon is the closest celestial body to the Earth, and its stark world has fascinated observers throughout the ages. It can be viewed even with a small telescope, such as Galileo had. The first time we look at the Moon in this way is an unforgettable experience. Suddenly, we see it as a world, a globe, not merely a flat, shining disc. The edge of the Moon appears jagged and uneven against the dark void of the heavens, provoking thoughts and questions about its real nature, hidden from us for so long. What does it really look like, what is it made of, how big, how far is it? Many of these mysteries have now been revealed to us; after thousands of years of fascinated speculation, we know a great deal about our nearest neighbor. Some mysteries still remain, but one thing is for sure: there is more than just green cheese up there!

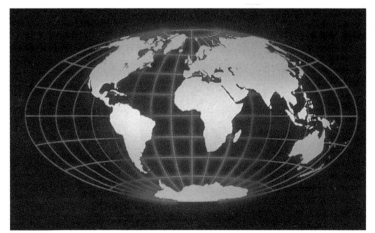

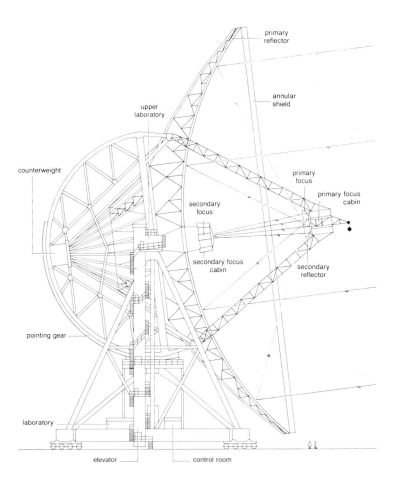

Enticing though the Moon may appear on a clear, starlit night, we should perhaps remind ourselves, as we gaze yearningly up, that it has none of the delights of Earth. It bakes in the sun and freezes in the shadow. The midday temperature at the equator reaches as high as 260 degrees F. And at midnight it can be cold beyond our imagining, at -273.2 degrees. Those delights that we take for granted on Earth – the infinite variety of life, terrain, vegetation – are completely missing on the Moon's barren, unchanging and inhospitable surface, because it has no atmosphere. So what is there on the Moon to entice us?

Above: The radio telescope showed the inhospitable appearance of the Moon's surface (right) in vivid detail. In this case looks are not deceptive, as the barren landscape swings between freezing and scalding.

As seen from Earth, the size of the Moon appears to change. Its orbit around us is elliptical, so that when it is at perigee (left) it is up to eleven per cent closer than at apogee (right).

Let us look at the Moon in relation to us, the Earth. The Moon is an average distance of 384,400 kilometers, or 238,7132 miles, from Earth.

That is close to thirty Earth diameters, or it's like traveling between New York, San Francisco and back again twenty-five times. To put it another way, the Moon is just over one light-second (the distance light travels in one second) from the Earth. That compares with galaxies which are several billion light years away from us. Or, alternatively, it would take just one-quarter of the population of China, standing on each other's shoulders (each adding 4 feet to the chain), to reach the Moon!

As you know, the Moon is our nearest neighbor. Venus, our closest planet, is 105 times further away than the Moon, and Neptune is never closer than 11,208 times the distance from Earth to the Moon. The Moon's distance from the Earth can be measured to within a certainty of one foot! This is thanks to special reflectors left on the Moon by the *Apollo 11* astronauts. They reflect laser light which can be measured as it travels between the Moon and the Earth.

The Moon and the Earth are not, however, always at the same distance from each other. There is a variation of as much as eleven per cent, due to the fact that the Moon orbits around us in an ellipse. When it is closest to us it is at "perigee", when it is most distant it is at "apogee". At perigee it can be as close as 221,456 miles (345,410 km) from Earth, at apogee as far away as 252,711 miles (406,700 km). So the distance between the Earth and the Moon varies by just over 30,000 miles – it depends whether it is on the "squashed" or the elongated part of its orbit.

Because the Moon is being pulled by both the Earth and the Sun, its orbit is elliptical.

Average orbital speed:
0.63 miles per second.

Surface area:
23,712,500 square miles.

Density:
3.34 times the density of water.

Brightness:
approximately one-millionth of that of the Sun.

Relative size:
smaller than any other planet in the Solar System.

Diameter:
2,160 miles (3,456 km) – that is just over a quarter of the diameter of the Earth.

Mass:
slightly less than one-eightieth of the Earth's.

Surface area:
less than one-thirteenth of that of the Earth.

Volume:
about one-fiftieth of that of the Earth.

Radius:
about 1,086 miles (1,738 km). That is, 0.2725 of that of Earth.

Surface gravity:
one-sixth of that of Earth.

Atmosphere:
none, therefore there are no clouds, there is no weather and no sound. There are, though, traces of hydrogen, helium, neon and argon.

Age:
4.6 billion years.

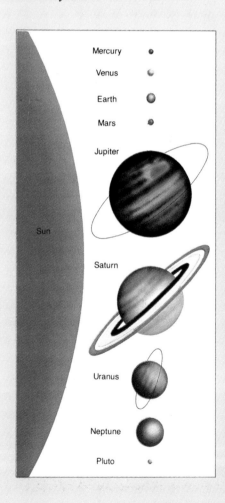

The Size of the Moon

How does the Moon compare with Earth in size? It weighs only just over one-tenth of the weight of the Earth – to be exact, eighty-one quintillion tons – or the combined weight of around 1,400 million, million, million, fifty-ton trucks, or twenty billion fully laden jumbo jets! Even this massive size is a mere 1.23 per cent of the size of the Earth.

The diameter of the Moon is just over 2,000 miles, or 3,476 kilometers, which is about a quarter of the Earth's 7,926 miles. That is about the distance between San Francisco and Cleveland. In fact, you could just about fit the Moon into America. Or, put another way, if the Earth were the size of your head, the Moon would be the size of a tennis ball.

Within the solar system, our Moon is only the sixth in order of size of moons, but in proportion to the size of the planet around which it revolves, it is by far the largest of the satellites. For example, if Jupiter (the largest planet) had a satellite proportionally as large as ours, it would be almost as big as Neptune. So the Earth and the Moon could be said to be a twin planet system, unique in the solar system. We only ever see one side of the Moon, because it spins on its axis, keeping the same side towards us and illuminated by its heavenly companion, the Sun. That is why we only ever see the same features in the same position, whatever phase the Moon is going through.

To picture how the Moon can only ever show one face to us, but still spin on its axis, try this. Imagine yourself as the Earth, with the Moon represented by a ball marked with a cross, in your outstretched hand. Turn round slowly, and the cross on the ball will remain visible to you as it makes a complete rotation round you. Yet the cross will have faced all four walls of the room in turn, and thus the Moon rotates in a month.

Above: The relative size of the Moon to the Earth.

Right: And yet how small the Earth may seem too . . . Apollo 12's view of our planet and a star from 30,000 miles out in space.

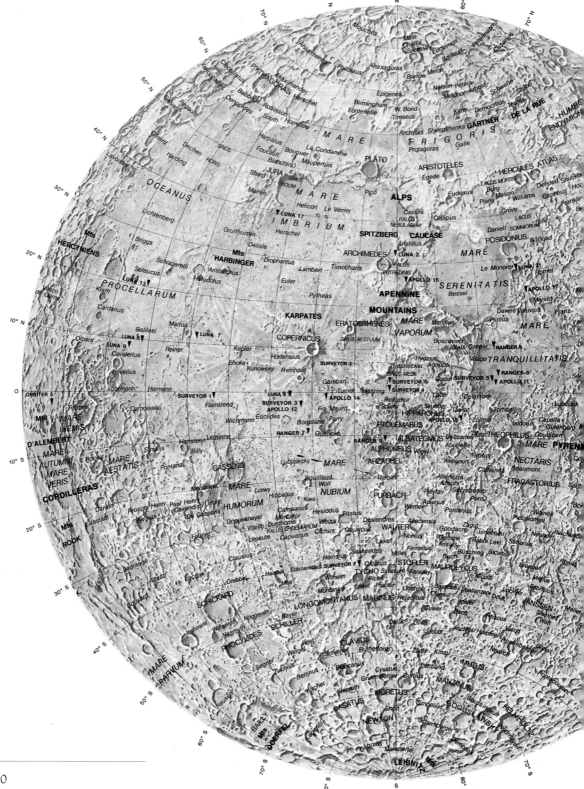

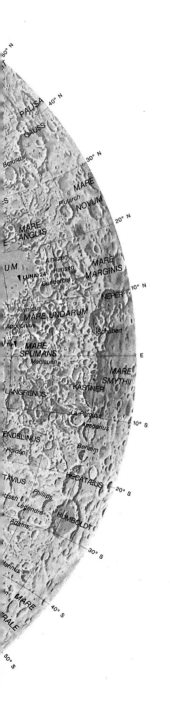

Chapter 2

The Origins of the Moon

Even today we do not know with absolute certainty how, and from where, our partner came to spin through the heavens with us, although we have a pretty good idea, particularly since the Apollo missions brought Moon rocks back for analysis.

The three main theories of the Moon's origins are:

1. The fission theory. This claims that shortly after the formation of Earth, it was spinning incredibly fast – one revolution about every two or three hours – and as a result, part of the surface was flattened. So violent was the spin that part of the equator broke off and spun into space, where it settled into orbit to become the Moon.

An appealing theory – it's rather romantic to imagine a part of us so far away, yet still linked – but unfortunately, almost certainly not true. Scientists have now shown that it just does not correlate with the way the Moon orbits, and that the Moon's rocks are chemically too different from the Earth's, and probably older.

2. The debris, or binary planet theory. This says that the Moon was formed at the same time as Earth from debris flying in space, which solidified into the Moon. This theory, too, although not completely discounted, has taken a hammering since the Moon missions. Again, the orbit doesn't tally, and for this theory to work, the two bodies should have the same densities, which they do not.

3. The capture theory. According to this, the Moon was formed elsewhere in the solar system, and became drawn into the Earth's gravity like a ball in a net. This is currently the favored explanation. But where did it come from? Analysis of lunar material suggests that the Moon may have originally been within the orbit of Mercury, the Sun's closest satellite. This also tallies with the fact that, although the law governing the distances of the planets from the Sun allows for a planet within this orbit, such a planet is "missing".

Perhaps, then, the Moon and Mercury moved close to the Sun in nearly circular orbits, the interaction of the two reaching such a degree that the Moon was moved into an elliptical orbit which carried it near the Earth, whereas Mercury, being four times the mass, settled into its present elliptical path.[140]

This may sound extraordinary, but the creation of moons is not an unusual phenomenon. There are forty-three moons in our own system, some planets having several.

And on the subject of the inconstant, the Moon and the Earth are both affecting each other. The Moon makes the Earth slow down. The flowing of water through tides, caused mainly by the Moon, produces friction on the ocean basins, and this is decreasing the rate of the Earth's rotation by 1 second every 50,000 years. That means that 400 million years ago the days were around twenty-two hours long instead of twenty-four. In another 35,000 years from now there will be twenty-five hours in a day – bliss for the workaholic!

Not only is the Earth slowing down, but as a result the Earth and the Moon are growing further apart. The Moon spirals away from the Earth at a rate of about three centimeters a year. The Moon was probably at its closest point to the Earth 1.2 billion years ago, approx. 20,000 kilometres (18,000 miles) away. It must have looked like a huge balloon, appearing twenty-two times its present size.

Unlike the Earth, the Moon has a very small core. This has been discovered by studying rocks from the Moon, which, in their composition, are quite unlike any known terrestrial rocks.

Crust (40-miles thick)

The structure of the Moon has been more positively determined since the manned landings carried out their investigations. It now seems likely that the Moon has a very small core compared with that of the Earth.

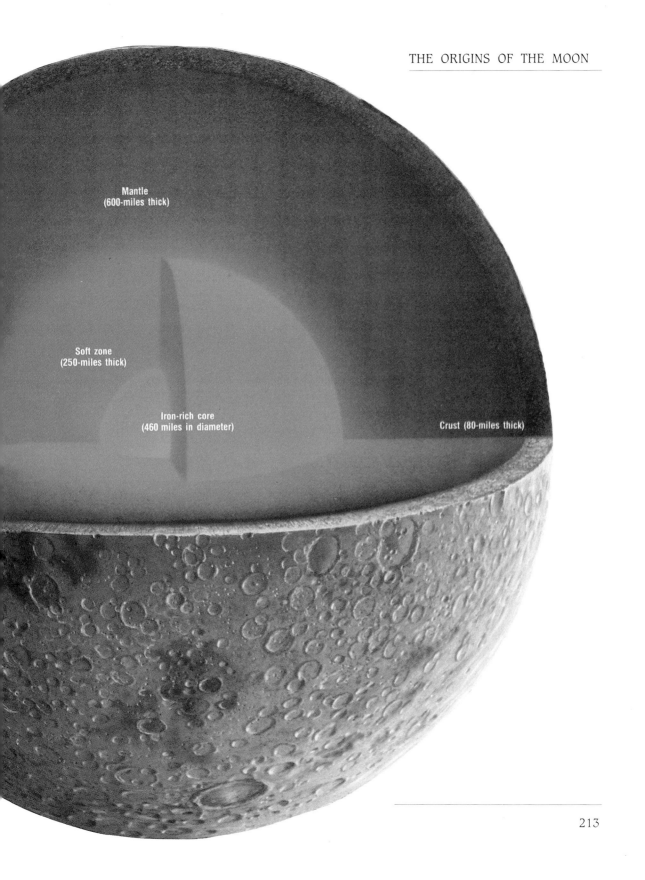

Mantle
(600-miles thick)

Soft zone
(250-miles thick)

Iron-rich core
(460 miles in diameter)

Crust (80-miles thick)

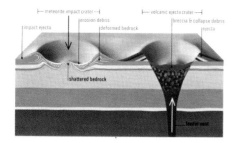

Rocks brought back by *Apollo* show that the Moon is 4,600 million years old, that is the same age as the Earth. But, unlike the Earth, what we see on the Moon remains unchanged by man. The Earth has changed radically since it was created, but most of the Moon's surface features were created within the first 1,000 million years of its existence. During this time it was subject to heavy bombardment by meteorites. Perhaps it is this feeling of timelessness that adds to the fascination that we, in our ever-changing environment, feel for the Moon. The reason for so little change on the Moon's surface is that there is very little atmosphere there. In fact, just the exhaust from *Apollo 11* deposited more gases than had previously existed.

Although the Moon has no atmosphere, occasional mist-like patches have long been observed on its surface. They mainly happen around the time of the perigee or apogee, and are probably connected with some kind of seismic activity – dust blown out by moonquakes.

As mentioned earlier, we generally believe that we only ever see half of the Moon's surface. This is not, strictly, true. Although the Moon does keep only one aspect turned towards us, it swings slightly, both from east to west and from north to south. About fifty-nine per cent of the Moon's surface is actually visible from Earth over a period of time.

The Moon reflects less than one-tenth of the light that falls on it from the Sun, although its brightness varies according to its phases because of the roughness of its surface and the consequent variable amount of shadow. The full moon – when Sun, Earth and Moon are in the same line – is ten times as bright as the Moon at first quarter.

Above: Craters were probably formed by meteoric impact and volcanic activity.

Below: The main features of the Moon's surface are mare, craters, faults and domes.

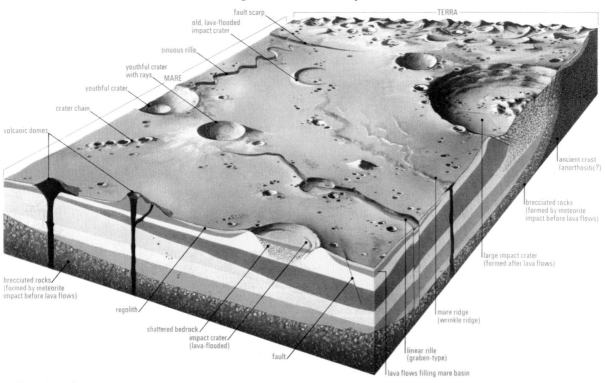

Gravity

What couples the Earth and the Moon? They both revolve around the center of mass of the Earth-Moon system. Think of it this way: imagine two people of very different weights on a see-saw. The heavier person must be nearer the center of gravity, or mass, for anything to happen.

And, as the Moon is eighty-one times further away from the center of mass than the Earth, it must have 1/81 times the mass.

So the Moon does not revolve around the center of the Earth! Each moves in an ellipse around the center of mass. The Earth-Moon center of mass lies approx. 3.5 miles (4,645 kilometres) from the Earth's center in the direction of the Moon.

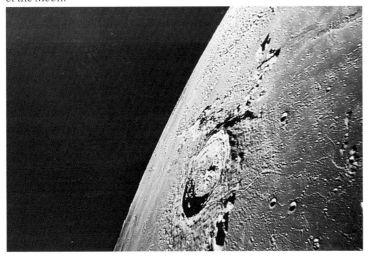

Above left: Eratosthenes, one of the largest craters in the western section of the Moon, has a central peak 61 kilometers in diameter.

The Moon is not a perfect sphere. The Earth's gravity pulls on it, producing an egg-like bulge along the main axis, which points at the Earth. This also has an effect on its rotation. Travel to the Moon and lose weight. Things weigh one-sixth of their Earth weight on the Moon, because the Moon has a smaller mass and radius, and so its surface gravity is one-sixth of that of the Earth.

This means that if you can jump three feet on Earth, on the Moon the same jump would reach a staggering eighteen feet, nine inches, while on Mars it would be seven feet, ten inches, and on Jupiter a mere one foot, three and a half inches, dropping to just one and a quarter inches on the Sun.

If the world record for an overhead lift is 564 and a quarter pounds, on the Moon the same effort would result in lifting 3,526 and a half pounds – that is the equivalent of two small cars. But an astronaut carrying a life-support system is still bearing the same mass as he would on Earth, and it takes the same effort to, for instance, stop running. It is a similar feeling to being under water, a floating in a kind of weightlessness. That is why astronauts appear so clumsy and slow.

Phases

"The moon's an arrant Theefe and her pale fire she snatches from the Sunne." [141]

The Moon makes no light of its own; all the light we see coming from it is reflected from the Sun. Half of the Moon is always in sunlight, but the amount of the light half that we can see from Earth varies from day to day. There are twenty-nine days, twelve hours, forty-four minutes and three seconds between one new Moon and the next.

The increasing Moon is known as waxing, the slowly vanishing one as waning. There are several different methods of dividing up the phases – commonly, into either two, four or eight distinct periods.

two periods:
waxing, increasing, new, light
waning, decreasing, old, dark

four periods: (astrological)
0-90 degrees after conjunction (0 degrees on the full moon).
90-180 degrees.
180-270 degrees.
270-360 degrees.

or	*or*
first quarter	new moon
second quarter	first quarter
third quarter	full moon
fourth quarter	last quarter

eight periods	*or*
new moon	new moon
crescent	waxing crescent moon
first quarter	Half moon, first quarter
gibbous	Waxing gibbous moon
full moon	Full moon
disseminating	Waning gibbous moon
last quarter	Half Moon, last quarter
balsamic	Waning crescent moon.

The apparent changes in the Moon's shape result from its changing position in relation to the Earth, and from the position of both of them in relation to the fixed direction of the Sun's rays. When the Moon is in nearly the same direction from the Earth as the Sun, its far side is in sunshine and we see at most a thin crescent visible just after sunset (the new moon).

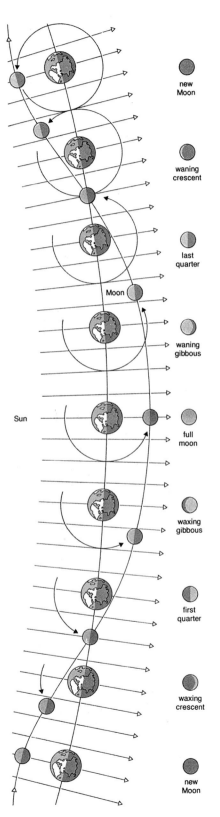

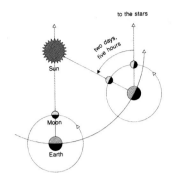

A week later, the Moon has moved a quarter of its way round the Earth, and the Sun illuminates half of the hidden hemisphere and half of the hemisphere turned towards Earth.

The first quarter begins when the Sun and Moon are cojunct, or in the same place, as viewed from Earth. At first the Moon cannot be seen, because it rises at the same time as the Sun. But by the end of the first quarter, a sliver of light is seen soon after sunset, as the Moon disappears – just after the Sun – over the western horizon.

The second quarter marks the time between the new moon and full moon. The right-hand half of the Moon is visible in the sky (the left-hand half if viewed from the southern hemisphere of Earth). At the beginning of the second quarter it rises around noon, and gives us light from dusk until it sets at midnight. At this time the Sun and Moon are at right angles (90 degrees square to each other).

A further week passes and the third quarter is marked at the beginning by the full moon. The Moon is almost directly opposite the Sun, and so is now fully illuminated. It is seen in the east from the time of rising at sunset to its setting at sunrise. It rises a little later each evening.

The fourth quarter is the stage half-way between full moon and new moon. The Sun and Moon are again at right-angles. The Moon now rises around midnight and is visible in the east in the last part of the night, until about sunrise. The illuminated disc has waned to a half-circle. A week later the month is complete and the whole cycle starts again.

The "line" separating the lit from the unlit part of the Moon is called the lunar terminator. It is elliptical, apart from the irregularities of the Moon's surface, because it is a circle seen in projection. It eventually becomes the full circle forming the edge of the Moon at new and full moon.

The Moon is called "triform" because of its different "faces"; either round, waxing with horns towards the east, or waning with horns towards the west. The Moon – like the Sun and the stars – appears to rise in the east and set in the west, even though its true movement relative to the north pole is from west to east. This is because the Earth spins west to east.

The Moon rises about fifty minutes later each day, because of its orbiting movement.

Left: The lunar cycle. The Sun, Moon and Earth change their relative positions during the course of the cycle.

Right: There are different ways of measuring a lunar cycle. The sidereal month is approximately two days shorter than a synodic month.

	New Moon (NM)	First Quarter (FQ)	Full Moon (FM)	Last Quarter (LQ)
aspect	●	◑	○	◐
position in the sky	near the Sun	at 90° away from the Sun	opposite the Sun	at 90° away from the Sun
rising	at dawn	at noon	at sunset	at midnight
setting	at sunset	at midnight	at dawn	at noon
times of visibility	invisible	late afternoon and evening	all night	second half of the night and early morning

Right: The ever-changing Moon shows us its many faces.

Bottom left and top right: new Moon.

Top left: eight-day old Moon.

Bottom right: crescent Moon with Earthsine.

Cycles

The Moon's regular waxing and waning is familiar to us all. But the Moon has a more hidden cycle too.

In the fifth century BC, a certain Meton of Athens calculated that the Moon's phases are repeated in a cycle of nineteen years, or 235 lunations. This has become known as the Metonic cycle of the Moon, or the lunar or minor cycle. This period of time has been considered of great significance: ancient Athenian public monuments were inscribed with letters of gold showing the dates of the full moon over the nineteen-year cycle. Each year had a "golden number" ascribed to it, according to its position in the Metonic cycle.

Meton did not get it absolutely right: and the Greek astrologer Callippus updated his work in the fourth century BC. He reckoned the precise cycle to be seventy-six years, and for good measure deducted a day at the end of it. This, he said, would bring the new and full moons to the same hour as well as day. But even this is not absolutely accurate – a whole day is lost every 553 years! The Metonic cycle has persisted, although it is now calculated at 18.61 years. It is accurate enough for most reckonings. In this time the Moon first rises and sets in extreme northerly and southerly positions, then moves inwards from these extreme positions for a period of 9.3 years, and outwards toward them again for a second 9.3 year period. The extreme positions reached by the Moon every 9.3 years are called its major and minor standstills. For a few days on either side of these standstills, a small perturbation, known as the Moon's "wobble", can be seen.

Once in every lunar cycle of 18.61 years, the Moon reaches its greatest height, as seen from the Earth. It reaches approximately this height once a month for over a year. This period is called the Moon's major standstill. A subsidiary effect of the major standstill is that, two weeks after its peak each month, the Moon rises only to a very low point in the sky. This means that if viewed from a northerly latitude the Moon hardly seems to set during the high trajectory, and during the low trajectory seems to move along just above the horizon.

The minor standstill occurs nine years after the major standstill in the 18.61-year cycle. At this time, the Moon appears for several months to reach the same low maximum height in the sky. (1) The nature of the Moon's motion is certainly complicated – and we have that on good authority; Newton himself once said that it was the only problem that made his head ache![142]

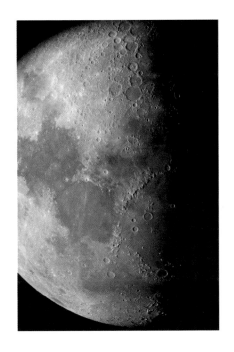

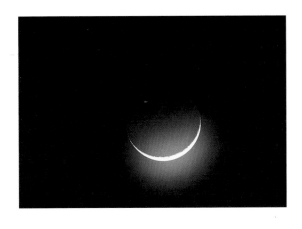

Tides

Moscow is raised and lowered by twenty inches twice a day by the tides, those mysterious currents that visibly affect the oceans of the Earth – and may affect much more besides, which in turn are affected directly by our friend, the Moon.

The Moon acts as a magnetic force on the Earth, pulling the waters towards it. The reason why Moscow, for instance, rises and falls in this extraordinary way is that the Earth acts as a giant sponge, absorbing and releasing the waters.

The general ebb and flow of tide-lore is that coasts have two low and two high tides daily, separated by an average of twelve hours and twenty-five minutes. The complete cycle of the tides takes about twenty-four hours and fifty minutes, as the Earth presents all its aspects to the Moon during that period.

What makes the Earth's waters ebb and flow? The Moon pulls gravitationally on both the Earth and the water. The Earth, being solid, is attracted to the Moon as if all of its mass was concentrated at a point in the center. But not so the water. It is free to move around, and the difference between the Moon's gravitational forces on the Earth and the ocean waters produces the tides. The tides are caused by the raising of two "bulges" in the oceans on opposite sides of the Earth. Water nearest the Moon is attracted more than the Earth itself which is, in turn, attracted more than the water on the other side of the Earth furthest from the Moon. The bulges do not "line up" with the Moon, but lag behind because of friction. This, as we talked about earlier, is gradually slowing the Earth down.

Some places, such as the Mediterranean, have no tides, and in other places there are many complex local variations to the tides. This is because the Earth is not a uniformly smooth ball, but is covered with things like continental shelves and oceanic basins, which have their own specific effects.

The Sun's attraction also contributes to tides, but with only a small percentage of the strength of the Moon. At times of new and full moon, the lunar and solar high tides add together, giving higher highs and lower lows. The highest, or spring tides, occur when the Moon and Sun pull along the same line, the lowest or neap when the Sun and Moon pull at right angles.

It takes about 1,500,000,000 horsepower for the Moon to drag the tides across the Earth each day. The Moon also pulls on the Earth's atmosphere as it does on the oceans. The air, like the water, ebbs and flows – the depth of air above is constantly changing the barometric pressure on Earth. Some parts of the Earth are more affected by this, as is the case with the oceanic tides.

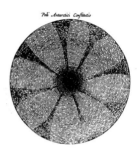

Just as the Moon raises a tide on Earth, so the Earth raises a tide in the solid body of the Moon. The tidal force causes a permanent distortion of the Moon – a bulge about 3,000 feet high in the direction of the Earth. This bulge, being closer to the Earth than the rest of the Moon, is more strongly attracted and therefore keeps pointing at the Earth.

The "tides" on the Moon have long since reduced the Moon's speed of rotation, until it has become synchronized with its revolution around the Earth – that is why the Moon keeps one face turned towards Earth, apart from a slight wobble. These tidal forces also mean that the Moon is gradually accelerating.

The Moon undergoes thousands of minor quakes each year, quite apart from impact by meteorites. Most of the quakes are triggered by the Earth's changing tidal pull (which of course is created by the Moon – 'which came first the chicken or the egg'). This pull is greatest when the Moon is at its closest point to the Earth. If the Moon can attract the Earth's oceans, what about smaller bodies of water? It has been proved, using highly sensitive equipment, that even a cup of tea is subject to lunar tides. (1)[143]

Left: The Moon pulls the Earth's waters like a magnet, creating tidal movement of the oceans.

Above: A popular 17th-century theory claimed that the seas well up in the south pole, creating great currents in the north.

Eclipses

"Darkness covered the Earth, and all the brightest stars shone forth. And it was possible to see the disc of the Sun, dull and unlit, and a dim and feeble glow like a narrow band shining in a circle around the edge of the disc. Gradually the Sun passing by the Moon (for the Moon was seen to be obstructing it in a straight line) sent forth its own rays and again filled the Earth with light." A total eclipse of the Sun, described by Leo Diaconus, AD 968

Eclipses have been held in awe throughout the ages, as a sign of the gods' displeasure, or an omen of some catastrophic event. Man has also managed to make use of them: the ancient Egyptian astronomers were able to predict eclipses from their knowledge of the eighteen-year eleven-day cycle, named the Saros. The priests kept the secrets of the Saros, which gave them apparent control over the heavens.

There are two kinds of eclipses, solar and lunar. A solar eclipse happens when the Moon passes between the Sun and the Earth, obscuring the Sun from our view. It causes a partial eclipse if the Earth passes through the Moon's outer shadow, or a total eclipse if the inner cone shadow crosses the Earth's surface (see diagram).

There is also what is known as an annular eclipse, which is when the Moon does not quite cover the Sun's whole disc, so that a ring of sunlight surrounds the darkened Moon. The area of the Sun which the Moon can cover differs because of the Moon's elliptical orbit around Earth, which means that, from the Earth, its size appears to vary. The annular eclipse happens when the Moon is slightly too far away to hide the Sun completely. Such eclipses are about as frequent as total eclipses, and in most years one of each kind will happen somewhere on the Earth's surface.

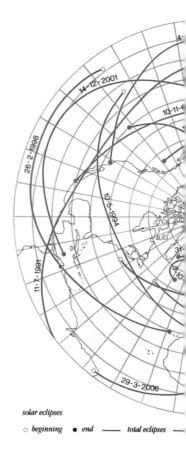

solar eclipses

○ *beginning*   ● *end*   —— *total eclipses*

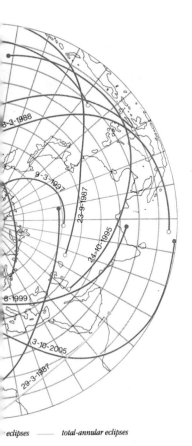

In a lunar eclipse, the Earth's shadow falls across the Moon when the Earth is between the Moon and the Sun, giving either a total or partial eclipse.

It is a curious coincidence (some might say no coincidence) that, although the Sun and the Moon are of such different sizes, their relative distances from the Earth are such that total eclipses of both bodies are possible. They appear to us to be virtually the same size, the Sun only slightly larger than the Moon. The Sun is, in fact, just over 400 times the size of the Moon, but is also on average 390 times further from Earth, which is why it usually only appears fractionally bigger than the Moon.

If the Moon orbited round the Earth in exactly the same way as the Earth travels around the Sun, there would be two eclipses each month – a solar one at the new moon and a lunar one at the full moon. The reason that this doesn't happen – and that eclipses are much rarer than this – is that the two planes of the orbits do not exactly coincide (see diagram). From one single spot on the Earth, it is unlikely that a total solar eclipse would be seen more than once in about 300 years. It has been estimated that if you stayed in the same spot on Earth for 1,000 years, you would probably see three total solar eclipses and four annular eclipses in that time. But since a total lunar eclipse is visible over half of the Earth's surface at once, most people will see several of these in a lifetime.

Top: The stages of a total lunar eclipse in September 1978, as seen from France. The Moon gradually emerges from the shadow of the Earth.

Left: Eclipse-spotting: The lines show the central points of total and annular solar eclipses in the northern hemisphere up to the year 2006.

eclipses ——— total-annular eclipses

A total eclipse of the Sun by the Moon can last for as long as seven minutes, fifty-eight seconds if it occurs in July, when the Sun is at apogee (furthest from Earth) and the Moon at perigee (nearest the Earth), and is observed from the Earth's equator. But the normal duration is much shorter than that.

When the Moon is eclipsed it appears to be a coppery-red color, quite different from its usual silver or deep gold.

There are never more than three eclipses of the Moon in a year, but there may be as many as seven eclipses a year, comprising either four solar and three lunar or five solar and two lunar.

Time

The monthly course of the Moon through the heavens has long been charted by man. The Moon could be said to be the earliest timepiece, and even now most of us are aware of the coming of a new or full moon. So how long is a lunar month? The answer to such a seemingly simple question is quite complicated.

There are several different ways of measuring the course of the Moon around the Earth (see table). The Moon orbits the Earth every 27.32 days. At least, it does according to one measurement, known as the sidereal month. Let's take a look at the different ways of measuring time by the Moon, according to whether it is judged against the Sun, the stars, the Earth's equator or the Earth's orbit.

Sidereal month:
27 days, 7 hours, 43minutes, 11.5 seconds

Synodic month:
29 days, 12 hours, 44 minutes, 3 seconds

Tropical month:
27 days, 7 hours, 43 minutes, 5 seconds

Anomalistic month:
27 days, 13 hours, 18 minutes, 33.2 seconds

1. Sidereal month: this is calculated using fixed stars as reference points. It is the true time the Moon takes to orbit the Earth. But in one sidereal month the Sun has moved eastwards in the celestial sphere by about 27 degrees, a distance which the Moon takes over two days to cover. So the total time the Moon takes to go from new to new, or full to full, overtaking the Sun at about 12 degrees per day, is twenty-nine and a half days, which makes a synodic month.

2. Synodic month: this is based on the phases of the Moon. There are just over twelve synodic months in a year – of the various different definitions,

this is the most natural kind of "month" (a word which also means "Moon"). The synodic month is also known as lunation.

3. Tropical month: this measurement is calculated from when the orbiting Moon passes through the plane of Earth's equator to when it next passes through the plane in the same direction. It is used by the International Astronomical Union. The tropical year is the time it takes the Sun takes to return to an exact position as measured for the equinox of date.

4. Anomalistic month: the time between perigee and perigee – when the Moon is closest to the Earth.

And as if all these different definitions were not sufficiently confusing, even the figures given for each of the types of months are only averages, because the motions of planets and satellites are not uniform. The movement of the Moon round the Earth and the Earth/Moon system round the Sun vary, making the whole topic of time-telling fraught with difficulties. In case that is not enough, here's yet another system: the points at which the Moon crosses the ecliptic (the Sun's apparent path among the stars during the year) into the northern and southern celestial hemispheres are called the ascending and descending nodes. The interval between crossings of a node, called the draconic month, is 27.212 days. It is thus called because of the ancient belief that eclipses were caused by the dragon having swallowed the Sun and Moon.

So time is flexible, it all depends on what system you use! There is a difference of more than two days in a month if the month is calculated according to the phases of the Moon (the synodic month) compared with a month measured by marking when the Moon returns to a particular point in the sky (the tropical month). Or, to put it another way, by the time the Moon has completed one revolution of the Earth relative to the stars, the Earth/Moon system has itself moved through 7.5 per cent of its orbit about the Sun. That is why the time between full moons, or new moons (called syzygies) is longer than a sidereal month, and is determined by using the Earth-Sun direction as a reference (see illustration).

Despite these complications, observation of the movements of Earth, Moon and Sun gave man the first means of measuring time. Even thousands of years ago, accurate lengths for the basic units of day, month and year were known, although not fully understood. The Moon particularly has been used to give time measurements between a day and a year. Its waxing and waning gives a continuous means of working out time, and is far more accurate than other natural phases of the movement of the stars.

The Moon's regular passage through the sky has meant that it has been used as one of the earliest means of measuring time.

The lunar month has normally been reckoned from the first sign of the crescent of the waxing Moon in the western evening sky. This is fine, until you use it for calculating years, which is when the trouble begins: twelve lunar months contain 354.4 days, or about eleven days fewer than a year as measured by the Sun. So the Sun calendar and the lunar calendar are out of synch.

Some attempts have been made to bring them together. The Babylonians added an extra lunar month to their calendar every two or three years.

The Greeks, Romans and Muslims adopted a year of twelve lunar months which alternately contained twenty-nine and thirty days. Eventually, Julius Caesar (around 45 BC) brought in the modern idea of dividing the year into twelve calendar months quite independent of lunations. Now the new moon can fall on any day of the month, not necessarily the first.

So the number of days in a year varies between cultures, and from year to year. Years with 365 or 366 days are based on the solar year, but the Muslim year is based on twelve lunar cycles, each of approximately twenty-nine and a half days, giving a total year of 354 or 355 days. The ancient Babylonian calendar was lunar – the month began when the crescent moon was first visible after sunset. For this reason, which seems strange to us today, the Babylonian day officially began in the evening. And at one time the Babylonians used a six-month calendar, based on lunar eclipses.

The Jewish year is also lunar. In ancient days, going as far back as the age of Saul, the time of the new moon was always celebrated. By the time Jerusalem became the capital of the Hebrew world, as soon as the first sliver of the new moon had been seen by credible witnesses, messengers would be dispatched to announce the beginning of a new month. The full moon was also considered very significant: the Passover was timed by the first full moon on or after the spring equinox. But nowadays, to keep roughly in line with the solar cycle, some years have twelve months (353, 354 or 355 days) and others thirteen (383, 384 or 385 days).

The ancient Egyptians calculated time according to changes in the Nile, the Sun and the stars, but they also had a lunar calendar to regulate festivals. They discovered that 309 lunations was almost equal to twenty-five civil years.

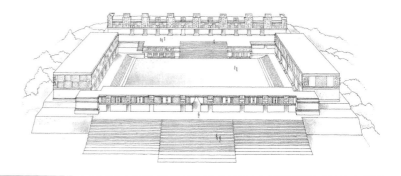

The ancient Mayas of Central America did not even have an invention such as the wheel, yet they worked out an amazingly sophisticated mathematical system in order to calculate their sacred days. The Moon was an important factor in this, and the results they came up with were astonishingly accurate. For example, the calculation of lunar months requires observations covering 405 full moons, or more than thirty-two years. The Mayas calculated that 405 full moons occurred in a period of 11,960 days. Today's astronomers have made the figure 11,959.88 days, so the Mayas were out only by one day every 292 years, or less than five minutes a year!

Because the lunar month is approximately twenty-eight days, this period became over time a convenient period for reckoning. This was probably especially so among nomadic peoples, who often journeyed during the cool of the night and thus depended on the Moon for their travel. The Christian world gradually adopted Caesar's method, but even now the Moon still creeps in to regulate time. Easter Day, the high point of the Christian calendar, is celebrated on the first Sunday after the first full moon after March 21, the time of the spring equinox, meaning that Easter can fall anywhere between March 23 and April 25.

The seven-day week, a shorter division of the twenty-eight day lunar month, is a remnant of the old lunar way of measuring time – and of course the number seven itself has taken on all sorts of esoteric significance.

So time is relative, it all depends on how you choose to judge it. And if we lived on the Moon, we would find that, because the Moon spins on its axis so much more slowly than the Earth does, each day – in terms of the Moon's rotation – would actually be equal to a month.

Bottom and top left: The ancient Mayan civilization made an impressively accurate calculation of lunar months, using their mathematical system.

Above: They even devised an accurate calendar for calculating religious festivals.

Surface

Even the naked eye can tell that the surface of the Moon is uneven. Dark patches form images from which myths and legends have sprung. These areas are the maria, or seas – so called by Galileo when he first turned his telescope Moon-wards. They contain no water, but the term has stuck.

The other main feature of the Moon's surface is the craters, which litter its surface and range in size from smaller than a coin to more than 150 miles (200 kilometers) in diameter. The craters have been the subject of much speculation – where did they come from? are they still forming? – and it is only in recent years, with the *Apollo* missions, that we have found out much about the Moon's barren surface. Most of the maria lie in the northern half of the side of the Moon which faces us. They are lower than the rest of the Moon's surface by around 2.5 miles (3 kilometers), which is why they appear darker. They are also known as the lunar lowlands, while the area surrounding them is the highlands. The maria are interconnected, and they have smoother surfaces than the brighter, cratered regions. We now know that they were formed by lava flow, and that this happened after the formation of the lunar crust and some of the craters.

The craters (named from the Greek word for "cup" or "bowl") proliferate on the Moon's surface. The largest one is about the same size as Connecticut. Although they vary widely in size, they have common characteristics: they are usually circular, they have small rims, and their floors are lower than the land outside the rim. In addition, the ground just around the craters has the appearance of having suffered an explosion.

How did they appear? There are two main theories: the impact theory claims that objects from space slammed into the Moon, forming the craters; the volcanic theory says they resulted from volcanic eruption. The latter theory became popular in the nineteenth century, but, for it to be correct, the Moon would need an interior hot enough to produce lava. We now know it does not have this.

So the impact theory is now the accepted one. Rays spreading out around craters suggest an impact unlike the result of volcanic eruption. Small particles orbiting the Sun have at some time come close enough to the Moon to be captured by its gravity. The fact that no large craters have been formed in recent times has given scientists further clues in thier quest to discover the age of the Moon.

The bleak, inhospitable surface of the Moon is covered with "seas" and craters, probably the result of impact by particles from space.

Far right: Craters vary in size from just a few inches across to several miles, like this one in the Sea of Tranquillity.

What does the lunar surface look and feel like? Surprisingly slippery. All the soil samples brought back by the *Apollo* missions had a high proportion of mainly round pieces of glass, which make the surface slippery.

These samples – undoubtedly the most expensive scientific samples ever gathered – yielded much new information about the nature of the Moon's surface. The top layer is porous, like a slightly sticky layer of debris, made up of fine particles (lunar soil) and larger pieces of rock.

The rock is of three main types: dark, fine-grained rocks made of magnesium/iron silicates; light-colored, grainy rocks called anorthosites, made of aluminium/calcium silicates; and rock and mineral fragments cemented together, called breccias. They have some characteristics which Earth rocks also have. But there is one major difference: while Earth rocks contain water in some of the minerals, Moon rocks are absolutely dry.

What has the study of these rocks told us about the Moon? They have shown that it was formed around 4.6 billion years ago. Once it had formed, the present highlands solidified – about four billion years ago. By comparing rocks from the different areas, we know that it was after this that the maria were formed, by laval flow.

As for the inside of the Moon . . . it is a peaceful place, compared with Earth! There are few moonquakes, and seismometers placed on the Moon by astronauts show that those which do occur are very minor, and happen deep down. This means that the Moon is cold and solid as deep as 800 – 1,000 kilometers beneath the surface.

What is beyond that? We do not know if there is a well-defined core, but the measurement of heat flowing up through the lunar surface suggests there is a lot of heat down there. It cannot be nickel-iron like the Earth's, because the Moon's magnetic field is too weak, but it is estimated that it could be as high as 1,500 K. The Moon still has a few secrets to reveal.

Chapter 3

Life on the Moon

We may have given up long ago the idea of finding Moonites and another civilization on the Moon, but why should there not be some form of life there?

The reason is that the Moon has too small a mass to retain an atmosphere suitable for life.

But what about life on the far side of the Moon? For years there was speculation that there might be landing places for extraterrestrial civilizations. This was all quashed when a Soviet spacecraft photographed the far side, and found it as barren as the front. Since man landed on the Moon in 1969, nearly eight hundredweight of lunar rock has been brought to Earth. The thing everyone was keenest to know was whether there was any trace of life of any sort.

The answer has been a resounding no. Over 3,000 different tests were carried out with the *Apollo 11* samples alone, but there were not even any advanced compounds of carbon and hydrogen. The minute traces of carbon that were present had probably arrived via meteorites and the solar wind. These disappointing discoveries led to the abandoning of quarantine for returning astronauts after the third Moon landing.

So it looks as if there is no life on the Moon. Well, not quite. In 1969 the *Apollo 12* astronauts brought back a camera taken from the *Surveyor 3* probe, which had landed automatically on the Moon two and a half years before. Scientists discovered streptococcus bacteria on it, which had survived all that time. So, despite the fact that the 120 deg. C daytime temperature drops to -180 deg.C during the lunar night, it seems that earthly bacteria can survive a considerable time on the Moon. There may be a bit of life in the old place after all.

Looking at the vibrant colors of a moon rock under a microscope, it's hard to believe that there's no life there. But tests have revealed there are no traces of advanced compounds of carbon or hydrogen in such samples.

Chapter 4

A Trip to the Moon

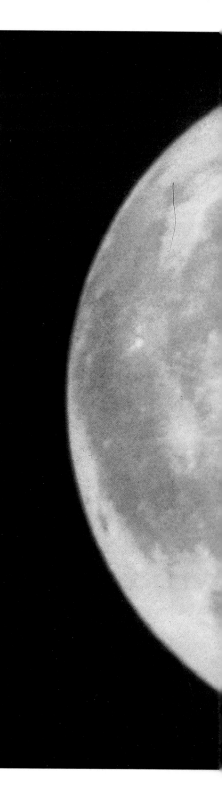

"America should commit itself to achieving the goal, before the decade is out, of landing a man on the Moon and returning him safely to the Earth."
(President Kennedy, 1961)

Between December 1968 and December 1972, twenty-four men went to the Moon. Of those, twelve landed and walked on its surface. No-one has been there since.

The first lunar landing took place at 3.18 p.m., Houston time, on 20 July 1969. The landing was watched back on Earth by about 600 million people – that was then about one-fifth of the world's population. The reaction on Earth was strangely mixed. President Nixon declared it "the greatest week in history since Creation". Others were more skeptical, seeing the £40 billion bill as outrageous in view of the Vietnam War and many social iniquities.

Yet others doubted that men *had* landed on the Moon. Even the NASA public affairs officer mischievously added to this doubt – he admitted that a film of astronauts training in a Michigan "moonscape" was indistinguishable from the real thing. Perhaps people just did not want their illusions shattered; they wanted to maintain the poetic image of the inviolable goddess above us.

> *"We shall send to the Moon, more than 240,000 miles from the control center in Houston, a giant rocket more than 300 ft. tall, made of new metal alloys, some of which have not yet been invented, capable of standing heat stresses several times more than have ever been experienced . . . on an untried mission to an unknown celestial body . . ."*
> (President Kennedy)

The story really began a hundred years before, when Jules Verne showed that the prospect of traveling to the Moon was a serious one. A few years later, in 1889, a Russian, Konstantin Tsiolkovsky, suggested that liquid-filled rockets could be used for such a purpose, and so began a century of exploration and discovery.

"As we set sail, we ask God's blessing on the most hazardous and dangerous and greatest adventure on which man has ever embarked."
(President Kennedy 1962)

Apollo 11 lifted off on 16 July 1969. Manned by three courageous men, Neil Armstrong, Edwin "Buzz" Aldrin and Michael Collins, heading off on the adventure of their lifetimes, they were fired by the enthusiasm and technological know-how of a planet that was rapidly growing smaller, as they were being thrust at a very high velocity on a direct route heading for that unknown entity, the Moon.

The two-and-a-half day, quarter-million-mile journey to the Moon went smoothly, almost belying the technological marvel that was taking place before our very eyes. Four mid-course correction maneuvers had been planned, but only one was needed.

Life on board the *Apollo* was an endurance test in its own right. The food consisted of packs of pre-frozen granules: "You couldn't tell what you were eating unless you read the menu", according to *Apollo 16's* Charley Duke. Disposal of human waste was even less savory: a bag was used, but as nothing will go into the bottom of a bag at zero gravity, the results were often disastrous! Washing was also extremely difficult, and the odor in the craft could hardly have been helped by the fact that the method of food preparation meant that a lot of flatulence was caused. Aldrin reports, "It got so bad it was

suggested we shut down our altitude-control thrusters and do the job ourselves!"

As the journey progressed, time lost its meaning for the crew, as Mike Collins noted. "Since humans generally define night as that time when our planet is between our eyes and the Sun, I suppose this must be considered daytime, but it sure looks like night out of several of my windows."

As for the weightlessness, most of the astronauts reported a euphoric, free feeling. Zero gravity had some other results, though, subtly re-sculpting their facial features. Probably the most disorientating thing was simply that there was no way to mark their progress. Alan Bean of *Apollo 12* noted: "After passing nothing and just essentially floating along and watching the Earth get smaller, all of a sudden you're at the Moon. That lack of way-points made it seem a little magical or mystical going there."

On 19 July *Apollo 11* disappeared behind the Moon, where the first of two critical lunar orbit insertion burns was made. This was like putting the brakes on – if it had not been successful, they could have either crash-landed on the back of the moon, or shot off into deep space. Ed Mitchell of *Apollo 14*: "For the first time in our entire flight we would be entirely on our own. We had to execute the burn correctly and come out on the other side of the Moon."

At Houston – and all over the world – the first signs of Apollo appearing over the lunar horizon were nervously awaited. When it did appear there was

When man launched himself towards the Moon, it was not just that world which he saw in a new light:
"Space travel has given us a new appreciation for the Earth. We realize that the Earth is special. We've seen it from afar, we've seen it from the distance of the Moon." (James Irwin, Apollo 15 astronaut)

great relief, but still some trepidation as communications were not immediately re-established due to a faulty antenna lock.

The astronauts, now no longer blinded by the Sun's glare, had for the first time had a good view of the Moon. The sight was stunning: it was back-lit by the Sun's corona as though in a solar eclipse. The Moon was lit by the glow of light from our own planet, "Earthshine". Mike Collins found it eerie: "A huge, three-dimensional sphere, almost ghostly, tinged sort of pale white. It was very, very large, and very stationary in our window, utterly silent, of course, and it gave one a feeling of foreboding." But other astronauts, such as Gene Cernan, felt as though it had "been waiting for us for millions of years".

Right: The "Eagle" has landed. The fragility of the tiny first lunar landing module is felt in the vastness of space.

Once in lunar orbit, the crew transmitted television pictures of the Moon's surface. The planned landing site, the Sea of Tranquillity, was spotted, and viewers around the world started becoming familiar with the names of lunar locations such as "Boot Hill", "Diamond Head Rille", and even "US Highway One". Emerging from the back of the Moon for the fourth time, the astronauts were treated to the spectacular sight of Earthrise, described by Mike Collins: "(The Earth) pokes its blue bonnet over the craggy rim (of the Moon) . . . and surges up over the horizon with an unexpected rush of color and motion. It is a welcome sight for several reasons: it is intrinsically beautiful; it contrasts sharply with the smallpox below; and it is home and voice for us."

The next stage was for Aldrin to enter the Lunar Module, check that everything was working and transfer equipment from the Command Module to the tiny landing craft. The inside was only about the size of two phone booths, with standing room only. They were reaching the moment everone had been waiting for – but first, some sleep was needed.

On waking, Armstrong and Aldrin transferred to the Lunar Module to make the final checks. The Lunar Module undocked, they were on their way. From now they adopted the call sign *Eagle* and the Command Module, which was left behind with Mike Collins aboard, became *Columbia*. This operation took place on the far side of the Moon.

The all clear was given from Houston after Collins had checked the Lunar Module for any damage. The descent engine was fired while the thirteenth orbit of the Moon was being made, the craft slowed down and the descent to the surface of the Moon began.

When the Command Module reappeared from behind the Moon, Collins reported, "Listen, babe, everything's going just swimmingly, beautiful".

Another two minutes, and the Eagle reappeared. Soon after that, another message from Collins: "*Eagle*, this is *Columbia*. They just gave you a go for PDI (Powered Descent Initiation)." In another five minutes, the *Eagle* had arrived at the "high gate", 50,000 feet above the Moon, about 260 miles from touchdown point.

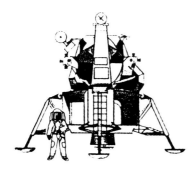

As the module approached the Moon it turned more upright, and a braking thrust was applied by firing the descent engine. This was the most critical part of the flight. The people directly involved and those hundreds of millions watching throughout the world felt the most excruciating tension.

When the *Eagle* was a mere five miles from touch down, Armstrong took over full manual control, a move that was necessary in order to avoid a shallow crater about the size of a football pitch, full of boulders.

Few people on Earth realized how close the *Eagle* came to disaster.

Steering away from the rocky area meant that they burnt up more fuel – Aldrin reckons he had only 10 seconds-worth left. In blissful ignorance of this, the world watched and listened:

"We're go! Hang tight. We're go."

"Seven hundred feet, twenty-one down, thirty-three degrees . . . Lights on. Down two and a half. Forward. Forward. Good. Forty feet, down two and a half. Picking up some dust . . ." Then those now famous words: "Houston, Tranquillity base here. The *Eagle* has landed."

Aldrin later claimed that he had spoken the first words from the Moon's surface: "Contact light!" But whatever was the case, man had landed on the Moon.

Right: Man on the Moon – Buzz Aldrin.

Far right: "One small step for man". Neil Armstrong descends from the lunar landing module onto the surface of the Moon, and into the history books.

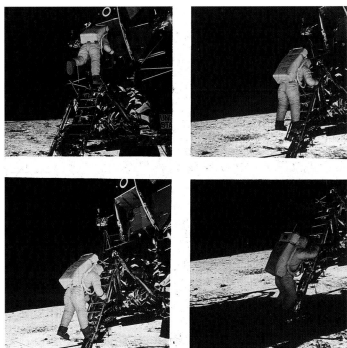

The view of the new world was described by Armstrong before he left the module: ". . . a relative plain cratered with a fairly large number of craters of five- to fifty-foot radius and ridges twenty, thirty feet high I would guess, and literally thousands of little one- and two-foot craters around the area . . . I'd say the color of the local surface is very comparable to what we observed from orbit . . . it's pretty much without color. It's gray and it's a very white chalk-grey . . . Some of the surface rocks in close here that have been fractured or disturbed by the rocket engine are coated with this light grey on the outside, but when they've been broken they display a dark, very dark grey interior." There was originally a sleep period scheduled for just after touchdown, but, hardly surprisingly, the astronauts were far too excited for this. So the first lunar walk was brought forward by four hours. Before this, Aldrin celebrated Holy Communion, a fact which NASA did not allow to be generally known, and Armstrong was noticing that the lunar surface looked very inviting – it appeared warm enough to sunbathe on! But instead of swimsuits, they put on space suits, the Portable Life Support System backpacks containing oxygen, a cooling system and a radio. Then the Lunar Module was depressurized.

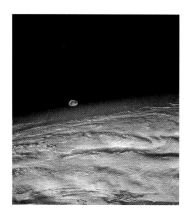

Armstrong reported that he was on the porch. A television camera attached to the outside of the Lunar Module (LM) transmitted the historic moment to Earth. He was seen to descend the ladder slowly.

"I'm at the foot of the ladder now," he said. "The LM footprints are only indented about one to two inches in very fine ground." And then, "That's one small step for man, one giant leap for mankind."

He described what he was seeing as he went. "The surface is very fine and powdery. It adheres like powdered charcoal to the soles of my boots . . . there seems no difficulty in moving about, as we suspected."

When Aldrin joined Armstrong, his comment was, "Beautiful. Beautiful . . . Magnificent desolation." Not to be outdone, Aldrin soon made his own "first" on the Moon. "My kidneys, which have never been of the strongest, sent me a message of distress. Neil might have been the first man to step on the Moon, but I was the first to pee in his pants on the Moon."

They proceeded to take samples and move about in the one-sixth gravity. Experiments were set up, including a laser ranging reflector for measuring the distance between Earth and the Moon, a passive seismic experiment package for measuring meteoric impacts and moonquakes, and a solar wind experiment consisting of a thin sheet of aluminium foil hung out in the lunar vacuum to catch any particles emitted by the Sun. Aldrin set up the American flag – specially designed to "fly" in the lunar vacuum but which proved difficult to jam into the Moon's surface. The first men to land on the Moon left behind this inscription: "Here men from the planet Earth first set foot upon the Moon. July 1969 AD. We came in peace for all mankind."

Back on Earth, President Nixon congratulated the two astronauts, apparently forgetting poor Mike Collins left alone on the Columbia. In fact, his often forgotten role was at least as daring as that of his comrades, as he orbited alone around the Moon.

Armstrong had been on the surface of the Moon for about two and a quarter hours when the command to return to the module came from Houston, who were becoming anxious about supplies in the backpacks. Armstrong and Aldrin returned to the *Eagle*, and completed the critical lift-off without a hitch, smoothly docking with the Command Module.

The two astronauts climbed through the docking tunnel to rejoin Collins, bringing with them boxes of lunar rocks and exposed film magazines. Then the *Eagle* was jettisoned, to stay in lunar orbit. The long journey home was a relaxed affair after the lunar landing.

"There's one sight I'll never forget," Armstrong said later. "As I stood on the Sea of Tranquillity and looked up at the Earth, my impression was of the importance of that small, fragile, remote blue planet."

At 400,000 feet above Earth the re-entry maneuver began, 1,500 miles from the intended splashdown point. The Service Module had been jettisoned, leaving the astronauts in the tiny, twelve-foot high, six-ton Command Module – all that was left of the original 363 foot, 3,000-ton original structure that had left Cape Kennedy eight days before.

The module approached Earth at over 24,000 miles per hour, protected by a heat shield. It had to enter a narrow corridor with absolute accuracy; if it was too high it would skip off the atmosphere and disappear into space; if it was too low the heat shield would burn away, taking the crew with it.

At 24,000 feet it had slowed down enough for the parachutes to be used to decelerate further. At 10,000 feet three main parachutes slowed it down to landing speed. Splashdown was within one mile of the target point, and within ten seconds of the time planned before the craft had left Earth.

A postscript on post-Moon life for the three astronauts: Neil Armstrong became a professor of engineering at the University of Cincinnati, shunning the media and commenting, "I just want to be a university professor and be permitted to do my research". Buzz Aldrin has suffered from occasional depression; as Mike Collins said, "Being an astronaut is a tough act to follow". Mike wrote a book, *Carrying the Fire*, in which he says: "The Earth continues to turn on its axis, and I am less impressed by my own disturbance to that serene motion, or by that of my fellow man."

The remaining *Apollo* missions could never have the same thrill of that first one, but they each had their own incredible achievement. The *Apollo 12* landing was calculated so precisely that the astronauts were able to bring back parts of the Surveyor Probe from the Moon. *Apollo 13* was scientifically a disaster, but was also the mission that made us most aware of the sheer human courage of the enterprise. Some 178,000 miles from Earth, an explosion disabled the service module.

Many experiments were made by the astronauts who set foot on the Moon, but the greatest experiment was the very journey to that other world.

The landing was abandoned; all effort went into bringing the three astronauts safely home. Miracles of skill and ingenuity were performed, and the command module *Odyssey* splashed down safely. *Apollo 15* was a spectacular mission. For the first time it used the Lunar Rover and drove to the great Hadley Rill, a huge crevasse. *Apollo 17*, in December 1972, took the last men to the Moon. When commander Eugene Cernan took the last step on lunar soil, an extraordinary era came to an end.

The quest to reach the Moon really can have no justification other than the old one of "because it was there". Or, as Neil Armstrong put it: "It's by the nature of his man's deep inner soul. Yes, we're required to do these things just as salmon swim upstream." But many spin-offs of space research, such as the development of computers, transistors, color TV, integrated circuits and lightweight plastics, have affected the way we all live.

The age of manned Moon-exploration has come to a close, at least for the foreseeable future. Perhaps we are still too close to it to realize its full significance, just as the import of Columbus' discovery was not seen in his lifetime. Let us give the final words to some of the men who went there, and for whom life will never again be the same. For they all have one thing in common: they have seen the wonder of our own planet from that vast distance.

> *"Space travel has given us a new appreciation for the Earth. We realize that the Earth is special. We've seen it from afar, we've seen it from the distance of the Moon. We realize that the Earth is the only natural home for men that we know of, and that we had better protect it."*
> (James Irwin, *Apollo 15* pilot)

> *"The Earth looked so blue and so round, and so small, so delicate. A house of all people that must be defended from harm."*
> (Alexei Leonov, first man to walk in space).

> *"I would have wished that after my return people had asked me how it was out there. How I coped with the glistening blackness of the world and how I felt being a star that circled the Earth."*
> (Reinhard Furrer, 1985 shuttle mission).

> *"Oh, I have slipped the surly bonds of earth*
> *And danced the skies on laughter-silvered wings;*
> *Sunward I've climbed, and joined the tumbling mirth*
> *Of sun-split clouds – and done a hundred things*
> *You have not dreamed of . . .*
> *And, while with silent, lifting mind I've trod*
> *The high untrespassed sanctity of space,*
> *Put out my hand, and touched the face of God."* [144]

APPENDIX

A Chronology of Space Travel

1924   The first serious technical study of rocket principles was published, by Hermann Oberth.

1926   A small liquid-filled rocket was launched by the American Robert H Goddard, reaching a height of 184 feet in 2.5 seconds.

1927   The German Society of Space Travel was formed to research rockets.

1943   The German V-2 rocket was launched in the Baltic, and traveled 122 miles (196 kilometers).

1949   Cape Canaveral, later called Cape Kennedy, was established by the Americans as a launching site in Florida.

1957   *Sputnik 1*, the first artificial Earth satellite was launched by the Russians on 4 October. This marked the beginning of the space age. *Sputnik 1* transmitted for twenty-one days, and stayed in orbit until **1958**. *Sputnik 2* was launched on 3 November, with the dog Laika aboard becoming the first mammal in space. The craft was destroyed on re-entry in April 1958, but Laika had died long before that. 1958 *Explorer 1*, the Americans' first satellite, was launched on 31 January from Cape Canaveral.

1959   The Russian *Lunik 1* was launched, becoming the first probe to go near the Moon. On 12 September *Lunik 2* became the first space probe to crash-land on the Moon. In October *Lunik 3* sent back the first photos of the Moon's far side.

1960   *Lunik 1* became the first artificial satellite of the Sun, having missed the Moon. The first space object to be recovered from orbit was the American satellite *Discoverer 13* capsule, landing in the Pacific Ocean. The Soviet dogs Belka and Strelka became the first animals to be recovered from orbit, when they returned in the *Sputnik 5* capsule.

1961   The first man in space, Soviet astronaut Yuri Gagarin, made one orbit of Earth in *Vostok 1* on 12 April. The Americans followed on 5 May when the first American astronaut, Alan Shepard, made a sub-orbital flight at 116 miles (187 kilometers). A second Russian spaceman, Gherman Titov, made a seventeen-orbit, twenty-four-hour flight in Vostok 2 on 6 August. Also in 1961, the Apollo project was authorized by President Kennedy.

1962   John Glenn, the first American in orbit, went around the Earth three times in *Friendship 7* on February 20. *Ranger 4* was the first US craft to reach the Moon, on 26 April. The Russians launched the first Mars probe, but lost contact.

1963   The first woman in space was Soviet cosmonaut Valentina Teresh-kova, on 16 June.

1964   The American *Ranger 7* was the first craft to take high-resolution television pictures of the Moon, on 28 July. The Soviets launched *Voskhod 1* on 12 October, carrying three crew.

1965   The first space walk – Alexei Leonov "walked" for ten minutes on 18 March, from *Voskhod 2*. Also in March, Ranger 9 took television pictures of highland regions. The first manned *Gemini* test flight took place. Ed White made the first American space walk on 3 June, from *Gemini 4*, and the first space rendezvous took place, when *Gemini 6* came within one foot of *Gemini 7* on 16 December.

1966   *Luna 9* became the first craft to soft-land on the Moon and relay panoramic and close-up pictures from there to Earth, on 31 January. The first docking in space took place, when Neil Armstrong and Dave Scott docked with the unmanned *Agena* rocket in *Gemini 80*, on 16 March. On 30 May *Surveyor 1* was the first American probe to soft-land on the Moon's surface. It took photos which were relayed to Earth. In December *Luna 13* carried out a mission similar to that of *Luna 9*, but also tested the lunar soil hardness.

1967   The first space disasters occurred. On 27 January three US astronauts (Ed White, Gus Grissom, Roger Chaffee) were killed in an *Apollo 1* launch-pad fire, and Soviet astronaut Vladimir Komarov was killed on 24 April when *Soyuz 2* crashed on Earth following parachute failure. *Surveyor 3* was the first craft to dig trenches on the Moon, on 17 April; *Surveyor 5* was the first craft to analyse lunar mare soil, on 8 September; *Surveyor 6* carried out the same kind of work in November.

1968   *Surveyor 7* made the first soft landing in the highlands, on 7 January. The first recovery of an unmanned lunar probe, Soviet *Zond 5*, was on 21 September from the Indian Ocean. In October the first *Apollo* mission carrying a crew, *Apollo 7*, held a test flight in orbit around Earth. *Apollo 8* made the first manned lunar flight, consisting of ten orbits, with Frank Borman, James Lovell and William Anders aboard.

1969   The first docking of two manned spacecraft took place on 15 January, when *Soyuz 4* and *5* exchanged cosmonauts by space walk. Americans made the first manned flight of the lunar module in March with *Apollo 9*. In May *Apollo 10*, with three astronauts aboard, descended to within six miles of the Moon's surface. On 16 July *Apollo 11* made the first manned landing, at the Mare Tranquillitatis. Neil Armstrong was the first man to walk on the Moon, followed by Edwin "Buzz" Aldrin, while Michael Collins remained in the orbiting command module. The lunar module *Eagle* blasted off from the Moon on 21 July, and docked with the command module *Columbia*. A successful splashdown was made on 24 July in the Pacific, after which the astronauts went into a twenty-one-day quarantine. Soviet astronaut Valery Kubasov, in *Soyuz 6*, carried out the first welding of metals in space in October. On 14 November *Apollo 12* made the second Moon landing, this time in the Ocean of Storms, with astronauts Charles Conrad, Alan Bean and Richard Gordon making moonwalks.

1970 In September *Luna 16* became the first unmanned craft to bring lunar rocks back to Earth. Two months later *Luna 17*, landing on the Sea of Rains, was the first unmanned craft to use a roving vehicle – *Lunakhod 1*. The eight-wheeled vehicle was driven by remote control from Earth, with a three-second delay.

1971 *Apollo 14* made the third manned Moon landing, with astronauts Alan Shepard, Stuart Roosa and Edgar Mitchell in February. This was the first manned landing in a highland region, Fra Mauro. In July *Apollo 15* made the first manned landing in which a roving vehicle was used, at Hadley Rille. There were also the first live pictures of lunar module take-off, and the longest exploration (eighteen hours) of the Moon's surface. The crew of Soyuz 11 was killed when the craft lost pressurization on re-entry.

1972 The unmanned *Luna 20* took soil samples from a highland area in January. In April, *Apollo 16* made the fifth manned lunar landing, in the highland region of Descartes. Astronauts Charles Duke and John Young explored for twenty hours and fourteen minutes, and Thomas Mattingly made a space walk of one hour twenty-three minutes during the mission. *Apollo 17* made the sixth and last *Apollo* landing on the Moon in December, with astronauts Eugene Cernan, Ronald Evans and Harrison Schmitt. It landed at the Sea of Serenity, stayed for a record seventy-five hours and took 243 pounds (110 kilograms) of samples from the highlands and the adjacent valley.

1973 *Luna 21*, an unmanned craft, landed in Le Monnier in January.

1975 The first joint Soviet-American mission in space took place, when *Soyuz 19* and *Apollo 18* docked while orbiting Earth, in July.

1976 The Soviet spacecraft *Luna 24* soft-landed on the surface of the Moon in August, in the Sea of Crises.

ACKNOWLEDGEMENTS

This book only came to be through the love and help of many others. My thanks to:

My family, for all their support, especially my parents who gave me a name I have come to treasure.

Chetan, for his love.

Everyone at Labyrinth, for their faith and encouragement.

Osho, Lord of the Full Moon. His finger has pointed to the Moon for me.

BIBLIOGRAPHY

(1) John Milton, *Paradise Lost*, describing Galileo looking at the Moon.
(2) *The Man in the Moone*, ed. F.K. Pizor and T.A. Comp, Sidgwick and Jackson, London, 1971, p.127).
(3) In Verne's *Round the Moon* of 1876.
(4) In *From the Earth to the Moon*.
(5) H.G.Wells' *The First Men in the Moon* (1901)
(6) B.Branston, *Gods of the the North*, Thames and Hudson. London and New York, 1955
(7) Pecock, *The Repressor of Over Much Blaming of the Clergy*, c.1449
(8) Hall, c.1595
(9) Charles Leslie, *Anthropology of Folk Religion*, Vintage, New York, 1969.
(10) Jane C. Goodale, *Tiwi Wives*, University of Washington Press, 1971
(11) D. Amaury Talbot, *Woman's Mysteries of a Primitive People*, Frank Cass, London, 1968
(12) *Religion in Primitive Cultures*, Wilhelm Dupre, Hungary, 1971
(13) Erich Neumann, *The Great Mother*, Princeton University Press, New York, 1963
(14) M.Eliade, *The Myth of the Eternal Return*, Princeton University Press, 1974
(15) B.Branston, *Gods of the the North*, Thames and Hudson. London and New York, 1955
(16) A.Holmberg, *Nomads of the Long Bow*, Natural History Press, New York, 1969
(17) Asen Balikci, *The Netsilik Eskimo*, Natural History Press, New York, 1970
(18) Claude Levi-Strauss, *The Naked Man*, Jonathan Cape, London, 1981
(19) *Religion in Primitive Cultures*, Wilhelm Dupre, Hungary, 1971
(20) R.H. Lowie, *The Crow Indians*, Holt, Rinehart and Winston, New York, 1956
(21) H. Butcher, *Spirits and Power*, Oxford University Press, Cape Town, 1980
(22) J.Middleton (ed.), *Gods and Rituals* Natural History Press, New York, 1967
(23) Claude Levi-Strauss, *The Naked Man*, Jonathan Cape, London, 1981
(24) M.Eliade, *The Myth of the Eternal Return*, Princeton University Press, 1974
(25) H.C. King, *The World of the Moon*, Barrie and Rockliff, London, 1960
(26) Ben Jonson, *Cynthia's Revels*, c.1601
(27) I.Silverblatt, *Moon, Sun and Witches*, Princeton University Press, USA, 1987
(28) A.Bancroft, *Origins of the Sacred*
(29) J. W. Slaughter, quoted in P.Katzeff, *Moon Madness*, Citadel, USA, 1981
(30) Shakespeare, *The Merchant of Venice*, V, i
(31) John Keats, *Endymion*
(32) Robert Graves, *Greek Myths and Legends*, Cassell, London, 1960
(33) *The Children of Diana, or How the Fairies Were Born* Charles Leland
(34) Christopher Fry
(35) Sylvia Plath, *Childless Woman*
(36) Lyall Watson, *Supernature*
(37) John Pope, The Rape of the Lock
(38) Zen Master Dogen, *Moon in a Dewdrop*, Element Books, California, 1985
(39) Zen Master Dogen, *Direct Mind, Seeing the Moon, 16th Night*
(40) Zen Master Dogen, *On a Portrait of Myself*
(41) Osho, *No Water, No Moon*
(42) Osho, *Zen: The Path of Paradox*, Vol. 3
(43) P.D. Ouspensky, *The Fourth Way*, Routledge and Kegan Paul, London, 1957
(44) P.D. Ouspensky, *The Fourth Way*, Routledge and Kegan Paul, London, 1957
(45) Shakespeare, *Love's Labour's Lost*
(46) Walter Scott, *Rob Roy*
(47) Wilkins, *New World*, 1638
(48) Shakespeare, *Macbeth*
(49) Shakespeare, *Henry IV*, Part 1
(50) Keats, *Endymion*
(51) D. Valiente, *An ABC of Witchcraft*, Robert Hale, London, 1973

(52) D. Valiente, *An ABC of Witchcraft*, Robert Hale, London, 1973

(53) J.G.Frazer, *The Golden Bough*, Macmillan, London,1922

(54) Margot Adler, *Drawing Down the Moon*, Beacon Press, Boston, 1979

(55) *Aradia, or the Gospel of the Witches*

(56) *Aradia, or the Gospel of the Witches*

(57) Scott Cunningham, *Earth Power* Llewellyn Publications, Minnesota, 1986

(58) Scott Cunningham, *Magical Herbalism*, Llewellyn Publications, Minnesota, 1982

(59) Joseph Campbell, *The Masks of God*, Vol.4 *Creative Mythology*, Secker and Warburg, UK, 1968

(60) Shakespeare, *Othello*

(61) Walton Brooks McDaniel, quoted in P. Katzeff, *Moon Madness*, Citadel, USA, 1981

(62) Walton Brooks McDaniel, quoted in P. Katzeff, *Moon Madness*, Citadel, USA, 1981

(63) Ben Jonson, *Devil's an Ass*

(64) John Dryden, *Amphitryon*

(65) H.J. Eysenck and D. K. B. Nias (65) (*Astrology*, Penguin Books, London, 1985

(66) Lyall Watson, *Supernature*

(67) D. Valiente, *Witchcraft for Tomorrow*, Robert Hale, London, 1985

(68) A.Puharich, *Beyond Telepathy*, Darton, Longman and Todd, London 1962

(69) Wordsworth, "To the Moon"

(70) Sir Thomas More, *Utopia*

(71) Keith Thomas, *Religion and the Decline of Magic*, Keith Thomas. Weidenfeld & Nicolson, London, 1971

(72) A.Bancroft, *Origins of the Sacred*

(73) A.L. Basham, *The Wonder that was India*, Sidgwick and Jackson, London,1979

(74) A.I.Berglund, *Zulu Thought – Patterns and Symbolism*, Indiana University Press, USA, 1976

(75) E.E.Evans-Pritchard, *Nuer Religion*, Oxford University Press, New York and Oxford, 1956

(76) R.F. Fortune, *Sorcerers of Dobu*, Routledge and Kegan Paul, London 1969

(77) Claude Levie-Strauss, The Origins of Table Manners, Jonathan Cape, London 1978

(78) *Kodausha Encyclopedia of Japan*, Japan, 1983

(79) Levi-Strauss, *Origin of Table Manners*

(80) I.Karp and C.J.Bird (eds.), *Explorations in African Systems of Thought*, Indiana University Press, USA, 1980

(81) D. Zahan, *The Religion, Spirituality and Thought of Traditional Africa*, University of Chicago Press, Chicago and London, 1970

(82) Wayland D. Hand, *Magical Medicine*, University of California Press, USA, 1980

(83) Aubrey Burl, *Rings of Stone*, Frances Lincoln, London, 1979

(84) Anne Bancroft, *Origins of the Sacred*, Arkana, London, 1987

(85) R. Castleden, *The Stonehenge People*, Routledge and Kegan Paul, London and New York, 1987

(86) M.Brennan, *The Stars and the Stones*, Thames and Hudson, London, 1983

(87) J. and C. Bord, *The Secret Country*, Paladin, 1978

(88) P.Devereux and I.Thomson, *The Ley Hunter's Companion*, Thames and Hudson, London, 1979

(89) Mark Twain, *Pudd'nhead Wilson's Calendar*

(90) Robert A. Millikan,Nobel Prize winner, 1924

(91) Fred Getting, *Visions of the Occult*, Rider, London, 1987

(92) H. Blavatsky, *Isis Unveiled*, 1877

(93) Joni Mitchell, *Little Green*

(94) Lesley Gordon, *The Mystery and Magic of Trees and Flowers*, Webb and Bower, Exeter, 1985

(95) *Llewellyn's Moon Sign Book*, Llewellyn Publications, USA, 1988

(96) E. Maple, *The Secret Lore of Plants and Gardens*, Robert Hale, London, 1980

(97) *Llewellyn's Moon Sign Book*
(98) Dyer, *English Folk-Lore*,1878
(99) L.Watson, *Supernature*
(100) Anon. *Patrick Spence*, c.1550
(101) Philip Sydney, *Astrophel and Stella*, 1591
(102) Shakespeare, *Julius Caesar*
(103) Thomas Heywood, *A Woman Killed with Kindness*, 1607
(104) Laurens van der Post, *A Walk with a White Bushman* Chatto & Windus, London, 1986
(105) Ibid.
(106) Tomlinson, *Arago's Astron*, 1854
(107) Keats, *Epistle to My Brother George*
(108) Emerson, *History*, 1841
(109) *The Rubaiyat of Omar Khayyam*
(110) Walt Whitman, *Dirge for Two Veterans*
(111) Shakespeare, *The Merchant of Venice*
(112) Ibid.
(113) Longfellow, *The Golden Legend*
(114) Shakespeare, *Romeo and Juliet*
(115) Ibid.
(116) Shakespeare: *A Midsummer Night's Dream*
(117) Milton, *Paradise Lost*
(118) Ibid.
(119) William Allingham, "The Fairies", from *The Music Master*
(120) Tennyson, *St Agnes' Eve*
(121) Tennyson, *The Princess*
(122) Wordsworth, *Christmas Minstrelsy*
(123) Keats, *Ode to a Nightingale*
(124) Sylvia Plath *The Moon and the Yew Tree*
(125) Byron, *We'll Go No More A-Roving*
(126) Coleridge, *The Ancient Mariner*
(127) Ibid.
(128) Stefano Guazzo, *Civile Conversation*, 1574
(129) Christina Rossetti, *Is the Moon Tired*
(130) Yeats, *The Crazed Moon*
(131) Keats, *Endymion*
(132) Lord Houghton, *The Moon*
(133) Walter de la Mare, *Silver*
(134) Walter de la Mare, *The Listeners*
(135) *The Private Journal of Henri Frederic Amiel*
(136) *The Diary of Alice James*
(137) "The Galoshes of Fortune", *Hans Andersen's Fairy Tales*, Black Ltd, London, 1912
(138) Osho, *The Path of the Mystic*, Rebel Press. Germany
(139) Erasmus, *Adagia*, 1508
(140) *Contact with the Stars* ,R.Breuer and W.Freeman, Oxford 1978.
(141) Shakespeare, *Timon of Athens*
(142) A.Service and J.Bradbury, *Megaliths and their Mysteries*, Weidenfeld & Nicolson, London, 1979
(143) Lyall Watson, *Supernature*, Hodder and Stoughton, London, 1974
(144) John Gillespie Magee,Jr. *High Flight*. Magee was shot down in the Battle of Britain

INDEX

PICTURE ACKNOWLEDGEMENTS

Ancient Art and Architecture
    53, 54, 57, 58, 59, 64, 91, 99, 138

Pictor International
    10, 14, 19, 22, 27, 32, 39, 42, 50/51, 60, 63, 83, 96, 106, 114/115, 119,
    125, 126, 127, 135, 153, 158, 167, 186, 195, 198, 202, 208, 209,
    234/235, 241, 243

Radio Times Hulton Picture Library
    227

Science Photo Library
    21, 95, 122, 130, 147, 154, 163, 166, 182, 188, 191, 194, 206, 215,
    219, 220/221, 230